A SHORT INTRODUCTION TO
NUMERICAL ANALYSIS

BY

M. V. WILKES, F.R.S.

Professor of Computer Technology
in the University of Cambridge

CAMBRIDGE
AT THE UNIVERSITY PRESS
1971

Published by the Syndics of the Cambridge University Press
Bentley House, 200 Euston Road, London NW1 2DB
American Branch: 32 East 57th Street, New York, N.Y.10022

Library of Congress Catalogue Card Number: 66–23108

ISBN:
0 521 09412 7 paperback
0 521 06806 1 clothbound

First published 1966
Reprinted 1971

Printed in Great Britain
at the University Printing House, Cambridge
(Brooke Crutchley, University Printer)

CONTENTS

PREFACE

This book is based on a course of introductory lectures that I have given for a number of years in Cambridge. I hope that it will be useful not only to students but also more generally to those who need to make use of a digital computer for scientific and engineering purposes. I have endeavoured to give the subject a modern slant and to confine myself to essentials.

The longest chapter is that on interpolation. This is not because of the practical importance of interpolation as such (it is, in fact, a rare operation to perform in a digital computer) but because the idea of the interpolating polynomial is fundamental to the use of finite difference methods in numerical analysis generally. The chapter on interpolation should, therefore, be regarded as laying the theoretical foundations for what is to follow.

Systematic use is made of difference operators for deriving finite difference formulae, although alternative methods are given in the more important cases. Since the view taken is that finite difference formulae are only proved when the functions concerned are polynomials, expressions containing difference operators may be regarded as convenient abbreviations for finite expressions that could be written out in full. No elaborate theoretical justification of the use of such operators is therefore called for. The finite difference formulae once derived are, of course, applied at user's risk to functions which can only approximately be represented by polynomials.

I would like to express my indebtedness to many colleagues at Cambridge and elsewhere. I am especially grateful to Dr J. C. P. Miller from whom I have learned a lot over the years. I would like to thank him particularly for the interest that he has taken in this book and for his very generous help in checking the proofs.

M.V.W.

June 1966

1

THE ROLE OF NUMERICAL ANALYSIS IN SCIENCE AND ENGINEERING

In science and engineering, we are typically concerned with some particular aspect of the physical world, and this we investigate by making use of a mathematical model. The use of a model serves two purposes—it enables us to isolate the relevant aspects of a complex physical situation and it also enables us to specify with complete precision the problem to be solved. When the model has been established, the next step is to write down equations expressing the constraints and physical laws that apply. These equations may be simple algebraic equations; on the other hand they may be differential or integral equations.

The equations must now be solved and here a choice presents itself. One way is to proceed by the methods of conventional mathematical analysis, in which case we shall hope to obtain the solution in the form of a formula or a set of formulae. Inspection of this solution may then yield qualitative results of interest; for example, it may be observed that one quantity varies exponentially with respect to some other quantity, that some variable has only a second-order effect on the result, and so forth. If quantitative results are required, they may be obtained by substituting numerical values in the formulae.

The alternative procedure is to express the equations by means of numerical analysis in a form in which they can be solved by computation. This leads, of course, directly to quantitative results. However, if enough such results are obtained, then qualitative results may emerge; for example, it may appear that one quantity is proportional—to the accuracy of the computations—to another, or that changes to one variable have only a slight effect on the result.

The difficulty with conventional mathematical analysis lies in solving the equations. As every schoolboy knows, it is easier to write down equations than to solve them. There is, moreover, something almost capricious about mathematical analysis; in the case of one differential equation, for example, it may be possible to obtain a useful solution, whereas it may be quite impossible to do so in the case of another equation, not recognizably more complicated than the first, or indeed quite similar to it. Numerical analysis is much more general in its application and usually when solutions exist they can be computed; non-linearity, for example, is not a matter of great account. In another sense, however, the numerical approach is less general in that special cases only can be handled and the entire computational process has to be repeated for each new case in which a solution is desired. This was a fatal drawback in the days when computing was slow and tedious. The coming of digital computers has changed the whole situation; indeed the missing generality has, to a large degree, been supplied, since it is possible for the computer program to be written in such a way that it can be used to compute as many special cases as are required merely by running it repeatedly with different starting values.

The great advantage of the numerical approach is that it enables more realistic models to be treated. The extreme difficulty of obtaining solutions by conventional mathematical analysis has led in the past to the use of highly unrealistic models simply because they led to equations that could be solved. In fact, the applied mathematician has been engaged in a continual tussle with his conscience to decide how far he could go in the direction of distorting his model in order to make the equations tractable. The point is well made in the ancient jest about the examination question that began 'An elephant whose mass can be neglected...'.

The presentation just given is, of course, much over-simplified. In practice, a combination of conventional mathematical analysis and numerical analysis is likely to be used. Indeed, the substitution of numerical values in a formula is a numerical process. The day has long gone by when numerical calculation was regarded as a last resort to be used when all else had failed. It is now, in many situations, the preferred method; sometimes, even when solutions in closed form exist, it may be simpler to go back to the original equations and solve them numerically.

The objective a scientist has in mind when carrying through an extensive series of computations is not necessarily to obtain precise numerical results in a particular case. He may be much more concerned with obtaining a general understanding of some physical situation and in determining what are the important factors. A judiciously planned series of computations, in which carefully chosen special cases are studied, may help him to do these things. The special cases are chosen to exemplify with a minimum of complication the features that it is desired to study; the mathematical models used are typically built up from uniform spheres, semi-infinite planes, and similar furniture from the world of the applied mathematician. On the other hand, it is sometimes necessary to make calculations using mathematical models that correspond much more closely to real objects in the physical world. An obvious field in which this applies is design engineering, but the need also arises in geophysics, astronomy, and similar sciences.

Whatever the end for which the calculations are required, however, the computing problem is very much the same. It is unfortunately not true that if results are required to a low degree of precision the calculations can be done throughout to the same low degree of precision. Sometimes it is necessary to work to quite high accuracy in order to get an answer which is accurate to 5 %. Even if it is not necessary, it may be the simplest thing to do. Often the accuracy of working is determined not so much by the accuracy required in the result, as by the necessity of providing evidence that the required accuracy has been achieved.

2

ITERATION

This chapter is concerned with the solution of the equation

$$f(x) = 0$$

by iteration. We start with an approximation x_0 to the solution and apply to it a procedure which yields another approximation, x_1. This new approximation, normally a better one, is now used as a new x_0, and the process repeated. The iteration is said to converge if a stage is reached at which, to the accuracy considered, $x_1 = x_0$. The material of the chapter may be extended to cover the case in which x is a vector. Iterative methods are of great importance in numerical analysis.

Newton–Raphson method. This is a standard method, of very general application, for deriving an iterative formula for the solution of $f(x) = 0$.

Let x_0 be the initial approximation and ξ the (unknown) error. Then we have

$$f(x_0+\xi) = 0$$

therefore

$$f(x_0)+\xi f'(x_0)+O(\xi^2) = 0$$

or

$$\xi = -f(x_0)/f'(x_0)$$

approximately. Thus, an improved approximation is

$$x_1 = x_0-f(x_0)/f'(x_0)$$

Examples. (1) If we take $f(x) = x^2-N$ we obtain Newton's iterative formula for a square root, namely

$$x_1 = \tfrac{1}{2}(x_0+N/x_0)$$

(2) If we take $f(x) = (1/x)-N$ we obtain the following iterative formula

$$x_1 = x_0(2-Nx_0)$$

This is sometimes used, in a digital computer that has a multiplier but no divider, for computing reciprocals and hence for effecting the operation of division.

Graphically, the Newton–Raphson method consists in drawing the tangent to the curve $y = f(x)$ at $x = x_0$ and finding the point x_1 at which this tangent intersects the x-axis (see Fig. 1).

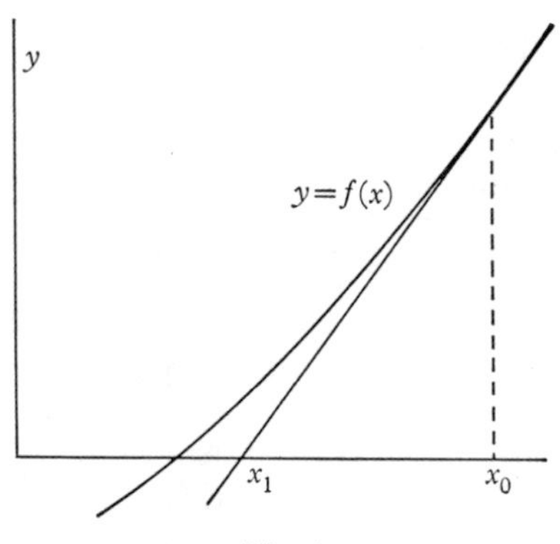

Fig. 1

The method could fail if the ordinate at x_1 failed to intersect the curve as in Fig. 2. In the case of an algebraic curve, however, there will always be an intersection if complex values are considered. This remark is not of theoretical interest only, since computer programs which handle such cases using complex arithmetic have been successfully written.

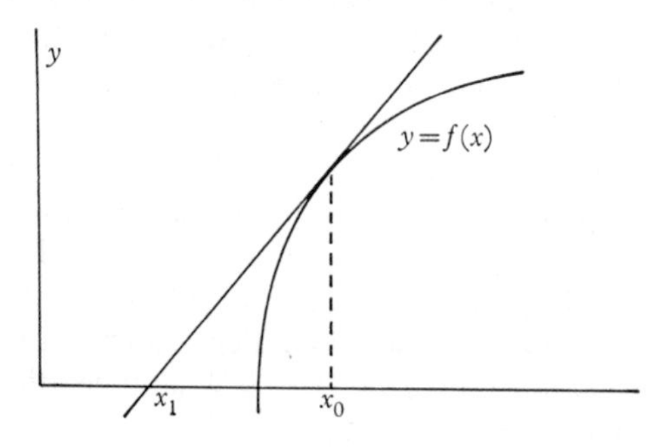

Fig. 2

If the computation of the derivative is lengthy, the same value may be used for several consecutive steps. Graphical considerations suggest that convergence will still take place; more steps may be required, but hopefully not enough more to outweigh the saving in time in calculating the derivative. This device—and others like it—was of more importance in the days of desk machine computing than it is now. It is mentioned here since reference will be made to it later in the discussion of convergence.

Example. To find to four decimal places the root of $x^3-3x = 3$ that lies near $x = 2$.

The Newton–Raphson formula gives

$$x_{n+1} = \frac{2x_n^3+3}{3(x_n^2-1)}$$

Numerical results are

$$x_0 = 2$$
$$x_1 = 2{\cdot}11$$
$$x_2 = 2{\cdot}10383$$
$$x_3 = 2{\cdot}10380\ 34032$$

Thus, to four decimal places, the root is 2·1038.

Notes. (1) In the above calculation, which was done on a desk machine, an increasing number of places was kept as the iteration proceeded and the accuracy improved. In a digital computer, one works, perforce, to the same accuracy throughout, except on the rare occasions when it is necessary to go to multi-precision working to get the required accuracy.

(2) A small computation error does not invalidate the result since the value obtained at the end of one step, even if there has been a small error, will still be a valid starting approximation for the next step. When convergence has occurred, one further application of the formula can be made to provide a check on the accuracy of the result.

(3) No theoretical proof of convergence has been offered, nor is it required; it suffices in a particular case that convergence is found to take place. (This does not mean that theoretical studies of the behaviour of specific iterative processes in model situations are not of great value as a guide to the choice of method to be used in practical cases.)

(4) Often iterative processes will converge from starting values very far removed from the final solution. For example, the process just illustrated converges from the value $x_0 = 4$, although this is much too far from the final value for ξ^2 to be negligible.

(5) Given a good starting approximation to a root, convergence will normally take place to that root. However, the approximation may not be as good as it appears to be, and in this case the process may converge ultimately to a more distant root. It is possible to construct cases in which convergence fails through the sequence of approximations becoming cyclic (e.g. $x_{n+2} = x_n$).

(6) Usually successive values of x_n are on the same side of the final value, except for the first which may be on the other side.

Other iterative formulae. The Newton–Raphson method, although quite general in application, is by no means the only way of obtaining an iterative formula. As an example of a formula derived in an *ad hoc* manner, consider again the equation

$$x^3-3x-3 = 0$$

which may be written $$x = (3x+3)^{\frac{1}{3}}$$

The fact that the right-hand side is less sensitive to a small change in x than the left-hand side suggests the following iterative formula:

$$x_{n+1} = (3x_n+3)^{\frac{1}{3}}$$

With the same starting value as before we find that

$x_0 = 2$	$x_4 = 2{\cdot}10353$
$x_1 = 2{\cdot}08$	$x_5 = 2{\cdot}10374$
$x_2 = 2{\cdot}0984$	$x_6 = 2{\cdot}10379$
$x_3 = 2{\cdot}1026$	$x_7 = 2{\cdot}10380$

Convergence does take place, but not very rapidly.

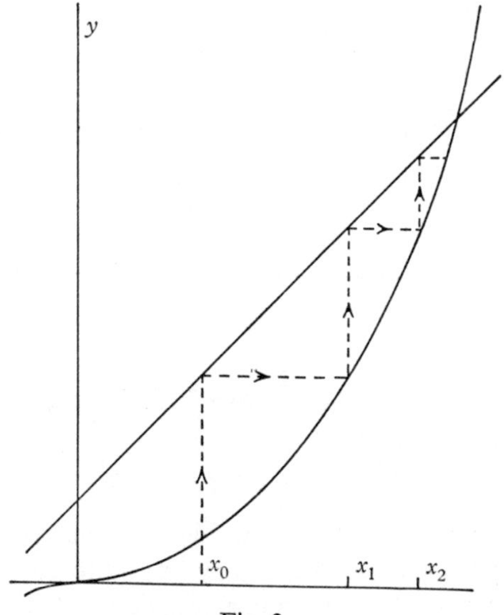

Fig. 3

Fig. 3 shows a geometrical interpretation of the process. The root of the equation is given by the intersection of the straight line

$$y = 3x+3 \tag{1}$$

and the cubic
$$y = x^3 \tag{2}$$
The first step of the iteration consists in finding the intersection of a line drawn vertically through x_0 with (1); this is equivalent to evaluating $3x_0+3$. A line is then drawn horizontally until it intersects (2); the abscissa is then x_1. Succeeding steps of the iteration take us along a kind of staircase to the intersection of (1) and (2). It is easily seen that if x_0 is taken to the right of the intersection, convergence still takes place, this time along a descending staircase.

In other situations, convergence may not take place independently of the starting point. For example in Fig. 4, there are two roots, a and b. It is easily seen that an iteration started with $a < x_0 < b$, or with $x_0 < a$ converges on to the root a. One started with $x_0 > b$ diverges to infinity. The root b is thus not accessible to an iteration of the type discussed.

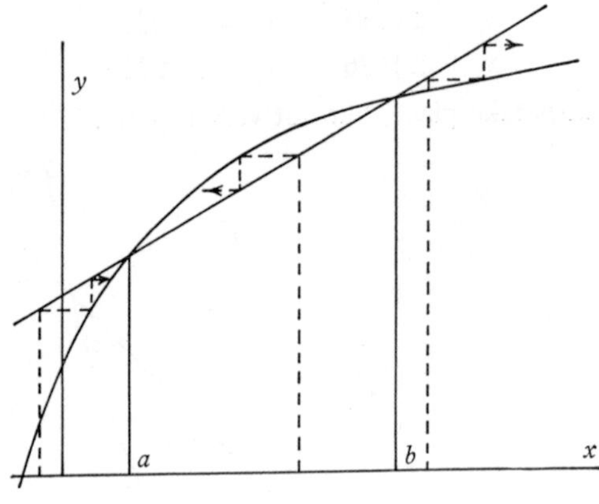

Fig. 4

Order of convergence. The iterative formulae considered above are special cases of the following:
$$x_{n+1} = F(x_n) \tag{1}$$
The reader will have noticed that, whereas the Newton–Raphson method for solving $x^3-3x-3 = 0$ may be said to go with a bang, the second method for solving the same equation just limps along. This suggests that the theoretical behaviour of an iteration based on (1) is worth investigating.

Suppose that x_n converges to the value X, so that we have
$$X = F(X)$$

If ξ_n is the error, assumed small, at any stage

$$x_n = X + \xi_n$$

so that

$$X + \xi_{n+1} = F(X + \xi_n)$$
$$= F(X) + a_1 \xi_n + a_2 \xi_n^2 + O(\xi_n^3)$$

where

$$a_1 = F'(X) \quad a_2 = \tfrac{1}{2}F''(X) \quad \text{etc.}$$

Therefore

$$\xi_{n+1} = a_1 \xi_n + a_2 \xi_n^2 + O(\xi_n^3)$$

Case I. $a_1 \neq 0$.

In this case convergence takes place if $|a_1| < 1$ provided that ξ_0 is not too large. If n is sufficiently large to make ξ_n^2 negligible, we have

$$\xi_{n+1} = a_1 \xi_n \quad \text{approximately}$$

and after a further $m-1$ steps

$$\xi_{n+m} = a_1^m \xi_n$$

This is *First Order Convergence* and may be described by saying that the number of extra steps required to obtain a given number of extra correct decimal places is independent of the number of steps taken. Thus if it takes five steps to go from 4 decimals to 6 decimals, it will also take five steps to go from 10 decimals to 12 decimals.

Case II. $a_1 = 0$, $a_2 \neq 0$.

Convergence now takes place whatever a_2 may be; it is sufficient that ξ_0 should be small. For large n

$$\xi_{n+1} = a_2 \xi_n^2 \quad \text{approximately}$$

or

$$a_2 \xi_{n+1} = (a_2 \xi_n)^2$$

This is *Second Order Convergence*, and $a_2 \xi_n$ is squared at each step. If $a_2 \sim 1$, the behaviour may be described by saying that the number of correct decimals is doubled at each step. Thus if one step improves the accuracy from 5 decimals to 10 decimals, the next step will improve it from 10 decimals to 20 decimals.

Case III. $a_1 = 0$, $a_2 = 0$, $a_3 \neq 0$.

This gives *Third Order Convergence*, and we may proceed similarly to define convergence of higher orders. The analysis is, of course, applicable only if $F(x)$ has a suitable number of derivatives.

In the second example given above

$$F(x) = (3x+3)^{\frac{1}{3}}$$

and

$$X^3 = 3X + 3$$

so that
$$a_1 = \frac{\partial F}{\partial X} = 1/X^2 \neq 0$$

Convergence is thus first order.

The following argument shows that, in general, the Newton–Raphson method gives second-order convergence. We have

$$F(x) = x - f(x)/f'(x)$$

therefore
$$F'(x) = f(x)f''(x)/[f'(x)]^2$$

Thus
$$a_1 = F'(X) = 0 \quad \text{since} \quad f(X) = 0$$

It is easily shown that, in general, $a_2 \neq 0$, so that the convergence is second order. The argument fails if $f'(X) = 0$, i.e. if $x = X$ is a multiple root; convergence is then first order.

The striking difference between the convergence noticed in the two examples given above is thus explained. Methods with second-order convergence are common, and are not limited to those obtained by any one method. For example, an alternative, also second order, to the usual Newtonian iteration for a square root is

$$x_1 = x_0(3N - x_0^2)/2N$$

Methods with third-order convergence, or better, are very rarely encountered in practice.

If, instead of computing $f'(x)$ afresh at each step, we make use over a number of steps of some (constant) approximation, g, to $f'(x)$ we have

$$F(x) = x - f(x)/g$$

so that
$$a_1 = F'(X) = 1 - f'(X)/g$$

which differs from zero unless g happens to be equal to $f'(X)$.

Thus it would appear that the penalty for approximating to $f'(x)$ is to make the process first order. Note, however, that the statement $a_1 = 0$ must be interpreted in numerical terms, i.e. as a statement true to the precision with which the work is being done. If $a_1 = 10^{-7}$ and we are working to 5 decimals, convergence will be second order. There is not an abrupt change from second-order behaviour to first-order behaviour as g is changed, but a gradual one.

Effect of rounding errors. Although the iterative process $x_{n+1} = F(x_n)$ may converge to a value X such that the equation $X = F(X)$ is satisfied to the number of decimal places considered, it does not follow that X is determined to the same number of places. In Fig. 5 the dotted curve represents $y = F(x)$ and the straight line represents $y = x$. The slope of the curve is near to unity, and it will be seen that there is a range of x for which $|x-F(x)| < \delta$. This range, which will be denoted by Δx, may be large compared with δ.

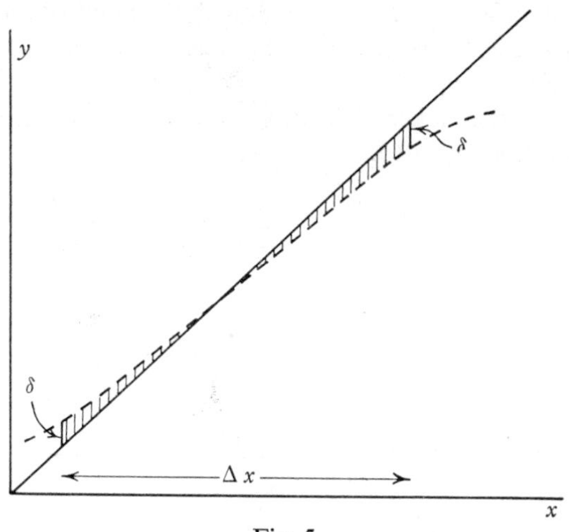

Fig. 5

An estimate for Δx may be obtained on the assumption that the curve $y = F(x)$ is effectively a straight line over the interval concerned. We then have

$$\delta = \tfrac{1}{2}\Delta x \left\{\frac{d}{dx}[x-F(x)]\right\}_{x=X}$$

or

$$\Delta x = \frac{2\delta}{1-a_1}$$

where, as before, $a_1 = F'(X)$. Thus if the iteration is carried to the stage at which $x-F(x)$ is zero to within the rounding error of half a unit in the last place retained, the error in X will be between $\pm\tfrac{1}{2}/(1-a_1)$ units.

If a_1 is nearly equal to 1, the error can be quite large. This is the case of first-order convergence, with the final value approached very slowly from one side. Since the final value is approached in this way, the error may be expected to have very nearly its maximum value, namely $\tfrac{1}{2}/(1-a_1)$. One should, therefore, be wary of iterations which converge very slowly from one side, since the value to which convergence takes place may differ by many units in the last place from the true value.

Successive inverse interpolation. This method is useful when computation of a derivative is inconvenient or impossible. Two starting approximations x_0 and x_1 are used, and a new approximation derived from the formula

$$x_2 = \frac{x_0 f(x_1) - x_1 f(x_0)}{f(x_1) - f(x_0)}$$

Geometrically, this is equivalent to finding the point of intersection of the x-axis and the line joining the two points on the curve corresponding to $x = x_0$ and $x = x_1$.

x_1 and x_2 are then taken to be new values for x_0 and x_1, respectively, and the operation repeated.

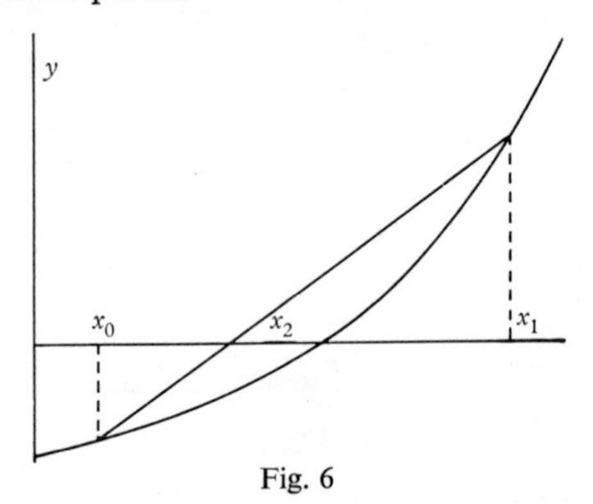

Fig. 6

Even if the root required lies between x_0 and x_1, the method can, if the initial approximations are not as good as could be desired, converge to another root outside that range. A variant of the method which prevents this, is to take for the new x_0 and x_1 the pair of values x_1 and x_2 if $f(x_1)$ and $f(x_2)$ are of opposite sign and x_0 and x_2 otherwise. The effect of this is always to work with a pair of approximations that straddle the required root. A disadvantage of this procedure is that convergence becomes very slow if, as often happens, the same approximation, x_r (say), is used repeatedly. A device due to Wheeler that substantially overcomes this objection is to use $\frac{1}{2}f(x_r)$ instead of $f(x_r)$ in the iterative formula whenever the approximation x_r is used repeatedly.

Examples

1. Find to 4 significant figures the two roots of $x = \log_{10} x + 2$, using Newton's method. Repeat using the iterations

$$x_{n+1} = \log_{10} x_n + 2$$

for the larger root, and

$$x_{n+1} = \tfrac{1}{100} \operatorname{antilog}_{10} x_n$$

for the smaller. [2·376, ·01024]

2. Use the following iterative formula for $1/\sqrt{a}$ to find $1/\sqrt{5}$ to four decimals

$$x_{n+1} = (3 - ax_n^2)x_n/2$$

Show that convergence is second order. [·4472]

3. An iteration is defined by $x_{n+1} = Ax_n/|x_n|$ where x_n is a vector, and A is a matrix. Take

$$A = \begin{pmatrix} 1 & 0 & 1 \\ 0 & 1 & 2 \\ 1 & 2 & 1 \end{pmatrix}$$

and x_0 a vector chosen at random, and show that to 3 decimals x_n converges to (1·023, 2·047, 2·288).

3

INTERPOLATION

The points in Fig. 1 suggest a continuous function of which intermediate values might be obtained by (for example) drawing a smooth curve.

In Fig. 2 no smooth curve is suggested. An experimentalist might draw a line as shown, and ascribe the discrepancies to experimental error. This is not our point of view in numerical analysis, where points are assumed to be known exactly, except for rounding errors.

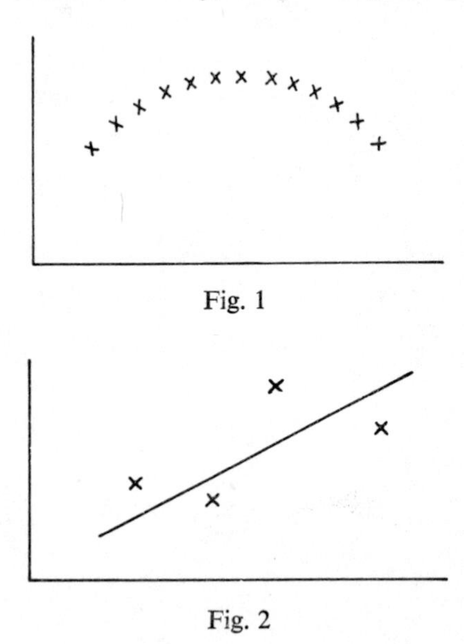

Fig. 1

Fig. 2

Interpolation is the name given to a process for obtaining intermediate values in the case of Fig. 1. It can be graphical or numerical; we are here concerned with the latter case. The points are then given in the form of a numerical table.

References to 'tables' include both published tables (e.g. sines, Bessel functions, etc.) and tables (perhaps consisting of a few entries only) computed during the course of some more extended calculation. A table may have only a fleeting existence in the memory of a digital computer.

The crudest method of interpolation is linear interpolation, familiar from the use of elementary tables. In linear interpolation it is supposed that the points are joined by a series of short straight lines; e.g. to find log 7·14

	Log	Difference
7·1	·8513	
		·0060
7·2	·8573	

$$\log 7{\cdot}14 = {\cdot}8513 + (4/10)\,{\cdot}0060 = {\cdot}8537$$

Algebraically, if the two points are (x_0, y_0), (x_1, y_1) we have for the interpolated value

$$y_\theta = (1-\theta)y_0 + \theta y_1$$

where θ = fraction of interval (4/10 in the above example). This may be written

$$y_\theta = l_0 y_0 + l_1 y_1$$

Lagrangean interpolation

The above method is equivalent to fitting a straight line

$$y = a_0 x + a_1$$

to a pair of adjacent points. In more elaborate methods of interpolation polynomials of higher degree are fitted (this is the numerical equivalent of drawing a smooth curve); e.g. for cubic interpolation we could fit

$$y = a_0 x^3 + a_1 x^2 + a_2 x + a_3$$

It is, however, more convenient to take the general equation of the cubic in the following form:

$$y = A(x-x_0)(x-x_1)(x-x_2) + B(x-x_{-1})(x-x_1)(x-x_2) + C(x-x_{-1})(x-x_0)(x-x_2) + D(x-x_{-1})(x-x_0)(x-x_1)$$

where (x_{-1}, y_{-1}) (x_0, y_0) (x_1, y_1) (x_2, y_2)

are the points through which the curve is to pass.
Putting $x = x_{-1}$, $y = y_{-1}$ we obtain

$$y_{-1} = A(x_{-1}-x_0)(x_{-1}-x_1)(x_{-1}-x_2)$$

This gives A; similar expressions are obtainable for B, C, D. Then

$$y = y_{-1}\frac{(x-x_0)(x-x_1)(x-x_2)}{(x_{-1}-x_0)(x_{-1}-x_1)(x_{-1}-x_2)} + y_0\frac{(x-x_{-1})(x-x_1)(x-x_2)}{(x_0-x_{-1})(x_0-x_1)(x_0-x_2)} + y_1\frac{(x-x_{-1})(x-x_0)(x-x_2)}{(x_1-x_{-1})(x_1-x_0)(x_1-x_2)} + y_2\frac{(x-x_{-1})(x-x_0)(x-x_1)}{(x_2-x_{-1})(x_2-x_0)(x_2-x_1)} \quad (1)$$

Normally we deal with tables with equal interval in x, i.e. with tables in which x_{-1}, x_0, x_1, x_2 are equally spaced. Put

$$x_{-1} = x_0 - h$$
$$x_0 = x_0$$
$$x_1 = x_0 + h$$
$$x_2 = x_0 + 2h$$
$$x = x_0 + \theta h$$

Then (1) may be written

$$y_\theta = L_{-1}y_{-1} + L_0 y_0 + L_1 y_1 + L_2 y_2$$

where
$$L_{-1} = -\tfrac{1}{6}\bar{\theta}(1-\bar{\theta}^2) \quad L_0 = \tfrac{1}{2}\bar{\theta}(1+\bar{\theta})(2-\bar{\theta})$$
$$L_2 = -\tfrac{1}{6}\theta(1-\theta^2) \quad L_1 = \tfrac{1}{2}\theta(1+\theta)(2-\theta)$$

where
$$\bar{\theta} = 1-\theta$$

In the normal applications of the formula θ lies between 0 and 1. This is Lagrange's 4-point formula; cf. Lagrange's 2-point formula given earlier:

$$y_\theta = l_0 y_0 + l_1 y_1$$

where
$$l_0 = \bar{\theta} \quad l_1 = \theta$$

Similar formulae are available for any number of points.

Higher order interpolation involves some work. With a digital computer this does not matter. In hand work labour is reduced by the availability of tables of the coefficients (L_{-1}, L_0, etc.). If one is prepared to use higher order interpolation, one can manage with a table with a much larger value of h and therefore of much smaller bulk.

Interpolation necessarily implies some loss of accuracy due to the accumulation of rounding errors; however, if the tabular values are accurate to within one half a unit in the last place, an interpolated value should very rarely be in error by more than one unit.

Example. To find $(3/2)^3$ from a table giving x^3 with $h = 1$. (A bad example, since one would not in practice interpolate if one knew that the function were actually a polynomial.)

x	y
0	0
1	1
2	8
3	27

Lagrange's 4-point formula with $\theta = \frac{1}{2}$ gives

$$y_{\frac{1}{2}} = \tfrac{1}{16}(-y_{-1}+9y_0+9y_1-y_2)$$

Then

$$y_{\frac{3}{2}} = \tfrac{1}{16}(-0+9+9\times 8-27)$$
$$= 27/8$$

Application of the Lagrange 4-point formula to the points of Fig. 2 would give a cubic of type shown in Fig. 3; this is useless unless it is known that the points really do lie on a cubic. Lagrange's formula must, therefore, not be applied blindly. It is a disadvantage of the formula that the user is given no warning of when its use is unjustified. We proceed to develop formulae that are more satisfactory in this respect, and for this purpose it is necessary to introduce difference notation.

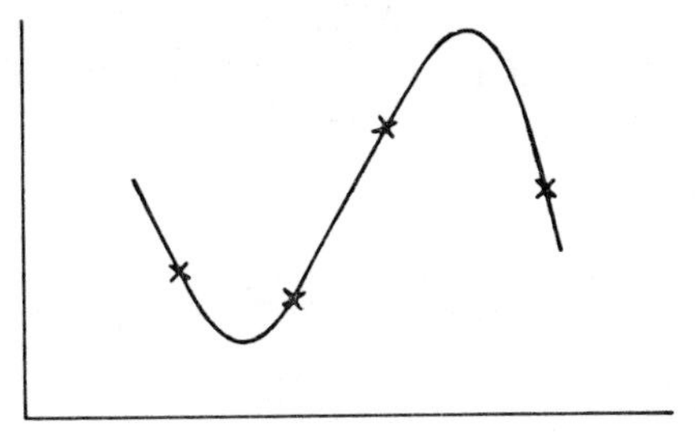

Fig. 3

Differences

Differences are best introduced by a numerical example

1		
	7	
8		12
	19	
27		18
	37	
64		

The numbers in the left-most column are given initially. The other numbers are *differences*, obtained by taking the two adjacent numbers in the column to the left and subtracting the upper from the lower.

There are three different notations for differences in current use. The most convenient notation to use depends on the application in hand.

Difference notation

$$
\begin{array}{llllll}
f_{-2} & & & & & \\
 & \delta f_{-\frac{3}{2}} & & & & \\
f_{-1} & & \delta^2 f_{-1} & & & \\
 & \delta f_{-\frac{1}{2}} & & \delta^3 f_{-\frac{1}{2}} & & \\
f_0 & & \delta^2 f_0 & & \delta^4 f_0 & \text{central differences} \\
 & \delta f_{\frac{1}{2}} & & \delta^3 f_{\frac{1}{2}} & & \\
f_1 & & \delta^2 f_1 & & & \\
 & \delta f_{\frac{3}{2}} & & & & \\
f_2 & & & & &
\end{array}
$$

$$
\begin{array}{llllll}
f_{-2} & & & & & \\
 & \Delta f_{-2} & & & & \\
f_{-1} & & \Delta^2 f_{-2} & & & \\
 & \Delta f_{-1} & & \Delta^3 f_{-2} & & \\
f_0 & & \Delta^2 f_{-1} & & \Delta^4 f_{-2} & \text{forward differences} \\
 & \Delta f_0 & & \Delta^3 f_{-1} & & \\
f_1 & & \Delta^2 f_0 & & & \\
 & \Delta f_1 & & & & \\
f_2 & & & & &
\end{array}
$$

$$
\begin{array}{llllll}
f_{-2} & & & & & \\
 & \nabla f_{-1} & & & & \\
f_{-1} & & \nabla^2 f_0 & & & \\
 & \nabla f_0 & & \nabla^3 f_1 & & \\
f_0 & & \nabla^2 f_1 & & \nabla^4 f_2 & \text{backward differences} \\
 & \nabla f_1 & & \nabla^3 f_2 & & \\
f_1 & & \nabla^2 f_2 & & & \\
 & \nabla f_2 & & & & \\
f_2 & & & & &
\end{array}
$$

In a numerical example the numbers would be the same in all three tables. Thus

$$f_1 - f_0 = \delta f_{\frac{1}{2}} = \Delta f_0 = \nabla f_1$$

$$\delta f_{\frac{1}{2}} - \delta f_{-\frac{1}{2}} = \delta^2 f_0 = \Delta^2 f_{-1} = \nabla^2 f_1$$

Note that

$$\delta^2 f_0 = f_{-1} - 2f_0 + f_1$$

δ, Δ, ∇ will later be regarded as operators; for the present, $\delta^2 f_0$ (for example) may be regarded as a composite symbol standing for a particular second difference.

The main difference between the three notations is in the way the suffixes are chosen. In central differences, suffixes are constant along a horizontal line, and this notation is, therefore, appropriate in the body of a table. At the beginning of a table, forward differences which have suffixes constant along a forward sloping line are convenient, and backward differences are convenient at the end of a table.

We will continue to write h for the interval in x, although it perhaps would be more consistent to write δx.

Differences of a polynomial. If a polynomial of degree n is tabulated at equal intervals in the argument, and a difference table formed, the nth differences will be found to be constant. That this is true may be seen as follows. Let

$$f(x) = a_0 x^n + a_1 x^{n-1} + a_2 x^{n-2} + \dots + a_n$$

a polynomial of degree n; then

$$\begin{aligned} \delta f_{\frac{1}{2}} &= f(x_0+h) - f(x_0) \\ &= a_0(x_0+h)^n + a_1(x_0+h)^{n-1} + \dots + a_n \\ &\quad - a_0 x_0^n - a_1 x_0^{n-1} - \dots - a_n \\ &= \text{a polynomial of degree } n-1 \end{aligned}$$

By repeated application of this argument it may be seen that the nth differences form a polynomial of degree zero, i.e. a constant.

Conversely, a polynomial may be constructed which has given constant nth differences.

Effect of a small error on a difference table. Note that there are errors of two kinds:

(1) rounding errors,

(2) mistakes.

The former must be expected to affect every value; the latter affect isolated values only.

Consider an isolated error, ϵ, in the function value f_0; we then have

$$\begin{array}{llll}
f_{-3} & & & \\
 & \delta f_{-\frac{5}{2}} & & \\
f_{-2} & & \delta^2 f_{-2} & \\
 & \delta f_{-\frac{3}{2}} & & \delta^3 f_{-\frac{3}{2}} + \epsilon \\
f_{-1} & & \delta^2 f_{-1} + \epsilon & \\
 & \delta f_{-\frac{1}{2}} + \epsilon & & \delta^3 f_{-\frac{1}{2}} - 3\epsilon \\
f_0 + \epsilon & & \delta^2 f_0 - 2\epsilon & \\
 & \delta f_{\frac{1}{2}} - \epsilon & & \delta^3 f_{\frac{1}{2}} + 3\epsilon \\
f_1 & & \delta^2 f_1 + \epsilon & \\
 & \delta f_{\frac{3}{2}} & & \delta^3 f_{\frac{3}{2}} - \epsilon \\
f_2 & & \delta^2 f_2 & \\
 & \delta f_{\frac{5}{2}} & & \\
f_3 & & &
\end{array}$$

Notes. (1) The resulting errors spread and increase in magnitude as the order of the differences is increased.

(2) The errors in each column have binomial coefficients.

The above remarks are illustrated by Tables 1 to 4.

TABLE 1

x	x^3	δ	δ^2	δ^3
2·5	15·625			
		4058		
2·7	19·683		648	
		4706		48
2·9	24·389		696	
		5402		48
3·1	29·791		744	
		6146		48
3·3	35·937		792	
		6938		48
3·5	42·875		840	
		7778		48
3·7	50·653		888	
		8666		48
3·9	59·319		936	
		9602		48
4·1	68·921		984	
		10586		48
4·3	79·507		1032	
		11618		48
4·5	91·125		1080	
		12698		48
4·7	103·823		1128	
		13826		48
4·9	117·649		1176	
		15002		
5·1	132·651			

Table 1. Exact values of x^3. The usual convention is followed of writing the differences as integers. Note that third differences are constant.

Table 2. The values of Table 1 rounded to two decimals; note that a number ending in 5 is rounded to be even. The third differences now fluctuate in a characteristic manner about a mean (actually 4·8). The fourth differences fluctuate about zero. Higher differences would be larger, and would also fluctuate about zero. Note that in this case the higher differences depend on the rounding errors *only*, not on the function.

TABLE 2

x	x^3	δ	δ^2	δ^3	δ^4
2·5	15·62				
		406			
2·7	19·68		65		
		471		4	
2·9	24·39		69		2
		540		6	
3·1	29·79		75		−2
		615		4	
3·3	35·94		79		0
		694		4	
3·5	42·88		83		3
		777		7	
3·7	50·65		90		−4
		867		3	
3·9	59·32		93		3
		960		6	
4·1	68·92		99		−3
		1059		3	
4·3	79·51		102		4
		1161		7	
4·5	91·12		109		−3
		1270		4	
4·7	103·82		113		0
		1383		4	
4·9	117·65		117		
		1500			
5·1	132·65				

Table 3. Extract from a table of $\log_{10} x$. The first ten values or so show fluctuations similar to those in Table 2; given only this piece of the table, one might conclude that one had a polynomial of degree three. The same applies to the last ten values or so, but on looking at the whole table, one sees that there is a gradual change in the mean value of the third differences. Over a small range, therefore, the function is indistinguishable from a third-degree polynomial, although it is not the same polynomial for all parts of the table. Thus we see, in a qualitative way, that the use of Lagrange's 4-point formula for interpolating this table is legitimate (e.g. log 4·175 is found to be ·620 6565, which is the correct value to within one unit).

Table 4. The same as Table 3, but with isolated errors deliberately introduced. Examination of the differences enables these errors to be detected and estimated. Note the occurrence of approximate multiples of the binomial coefficients (1 4 6 4 1) in the last column.

TABLE 3

x	Log x	δ	δ^2	δ^3	δ^4
4·00	·6020600				
		53950			
4·05	·6074550		−661		
		53289		14	
4·10	·6127839		−647		3
		52642		17	
4·15	·6180481		−630		−3
		52012		14	
4·20	·6232493		−616		2
		51396		16	
4·25	·6283889		−600		−4
		50796		12	
4·30	·6334685		−588		2
		50208		14	
4·35	·6384893		−574		−1
		49634		13	
4·40	·6434527		−561		0
		49073		13	
4·45	·6483600		−548		−1
		48525		12	
4·50	·6532125		−536		−1
		47989		11	
4·55	·6580114		−525		2
		47464		13	
4·60	·6627578		−512		−4
		46952		9	
4·65	·6674530		−503		2
		46449		11	
4·70	·6720979		−492		0
		45957		11	
4·75	·6766936		−481		−1
		45476		10	
4·80	·6812412		−471		0
		45005		10	
4·85	·6857417		−461		−2
		44544		8	
4·90	·6901961		−453		2
		44091		10	
4·95	·6946052		−443		−1
		43648		9	
5·00	·6989700		−434		−1
		43214		8	
5·05	·7032914		−426		0
		42788		8	

TABLE 3 (*cont.*)

x	Log x	δ	δ^2	δ^3	δ^4
5·10	·7075702		−418		1
		42370		9	
5·15	·7118072		−409		−1
		41961		8	
5·20	·7160033		−401		−1
		41560		7	
5·25	·7201593		−394		0
		41166		7	
5·30	·7242759		−387		1
		40779		8	
5·35	·7283538		−379		−2
		40400		6	
5·40	·7323938		−373		2
		40027		8	
5·45	·7363965		−365		
		39662			
5·50	·7403627				

TABLE 4

x	Log x	δ	δ^2	δ^3	δ^4
4·00	·6020600				
		53950			
4·05	·6074550		−661		
		53289		14	
4·10	·6127839		−647		103
		52642		117	
4·15	·6180481		−530		−403
		52112		−286	
4·20	·6232593†		−816		602
		51296		316	
4·25	·6283889		−500		−404
		50796		−88	
4·30	·6334685		−588		102
		50208		14	
4·35	·6384893		−574		−1
		49634		13	
4·40	·6434527		−561		0
		49073		13	
4·45	·6483600		−548		−1
		48525		12	

† Error = +·0000100.

TABLE 4 (*cont.*)

x	Log x	δ	δ^2	δ^3	δ^4
4·50	·6532125		−536		−1
		47989		11	
4·55	·6580114		−525		2
		47464		13	
4·60	·6627578		−512		−4
		46952		9	
4·65	·6674530		−503		2
		46449		11	
4·70	·6720979		−492		0
		45957		11	
4·75	·6766936		−481		−1
		45476		10	
4·80	·6812412		−471		0
		45005		10	
4·85	·6857417		−461		−2
		44544		8	
4·90	·6901961		−453		2
		44091		10	
4·95	·6946052		−443		−21
		43648		−11	
5·00	·6989700		−454		79
		43194		68	
5·05	·7032894†		−386		−120
		42808		−52	
5·10	·7075702		−438		81
		42370		29	
5·15	·7118072		−409		−21
		41961		8	
5·20	·7160033		−401		−1
		41560		7	
5·25	·7201593		−394		0
		41166		7	
5·30	·7242759		−387		1
		40779		8	
5·35	·7283538		−379		−2
		40400		6	
5·40	·7323938		−373		2
		40027		8	
5·45	·7363965		−365		
		39662			
5·50	·7403627				

† Error = −·0000020.

Interpolation using differences

Before systematic methods are developed for deriving finite difference formulae by means of difference operators, it will be shown how the Newton–Gregory interpolation formula may be obtained by an elementary method.

We start with the formula for linear interpolation

$$y_\theta = y_0 + \theta \Delta y_0 \qquad (1)$$

This is exact for a function of the form

$$y = a + bx$$

We apply it to the function

$$y = a + bx + cx^2$$

and calculate the error. Since the equations are linear, we may omit the first two terms, for which (1) is exact, and take simply

$$y = cx^2$$

For this function,

$$\Delta y_0 = c(x_0+h)^2 - cx_0^2 = 2chx_0 + ch^2$$

$$\Delta^2 y_0 = 2ch^2 \qquad (2)$$

The true value of y_θ is given by

$$y_\theta = c(x_0+\theta h)^2 = cx_0^2 + 2c\theta hx_0 + c\theta^2 h^2$$

The approximate value calculated from (1) is

$$y_\theta = cx_0^2 + \theta(2chx_0 + ch^2)$$
$$= cx_0^2 + 2c\theta hx_0 + c\theta h^2$$

The error is $-ch^2\theta(\theta-1)$; or using (2), $-\frac{1}{2}\theta(\theta-1)\Delta^2 y_0$.

We may use this error as a correction to (1), and obtain the following formula exact for a polynomial of the second degree

$$y_\theta = y_0 + \theta \Delta y_0 + \tfrac{1}{2}\theta(\theta-1)\Delta^2 y_0$$

Similarly, further terms may be obtained; the next one, making the formula exact for a cubic, is

$$\tfrac{1}{6}\theta(\theta-1)(\theta-2)\Delta^3 y_0$$

We have then for the complete Newton–Gregory formula

$$y_\theta = y_0+\theta\Delta y_0+\tfrac{1}{2}\theta(\theta-1)\Delta^2 y_0+\tfrac{1}{6}\theta(\theta-1)(\theta-2)\Delta^3 y_0+\ldots$$

Notes. (1) The first two terms $y_0+\theta\Delta y_0$ correspond to linear interpolation, and the other terms may be regarded as constituting a correction.

(2) For interpolation to be possible it is necessary that the terms should fall off in magnitude until they become negligible.

The above two remarks apply to other interpolation formulae expressed in differences, and the second in particular will be elaborated later.

The above formula terminates for a polynomial. To apply it to a function which is not a polynomial, taking terms as far as $\Delta^3 y_0$ (say), is equivalent to replacing the function by a polynomial through four points which, as the following table shows, are just enough to enable the third difference to be computed.

	y_0			
→		Δy_0		
	y_1		$\Delta^2 y_0$	
		Δy_1		$\Delta^3 y_0$
	y_2		$\Delta^2 y_1$	
		Δy_2		
	y_3			

The polynomial is identical to that fitted by Lagrange's 4-point formula.

The Newton–Gregory formula is convenient for use at the beginning of a table, with a value of θ between 0 and 1, giving interpolation in the first interval as indicated by the arrow. It could be used with θ greater than 1, but central difference formulae about to be developed are generally more convenient for use in the middle of a table.

One may not often need to work near the beginning of a table, but it it is very common to be working near the end of a table, since this is a situation that occurs whenever a table is being extended, for example, when the solution to a differential equation is being obtained by a step by step method (see Chapter 5). The Newton–Gregory backward difference formula is suitable for use in these circumstances. It may be obtained by a method similar to that used above and is as follows:

$$f_\theta = f_0+\theta\nabla f_0+\tfrac{1}{2}\theta(\theta+1)\nabla^2 f_0+\tfrac{1}{6}\theta(\theta+1)(\theta+2)\nabla^3 f_0+\ldots$$

If the formula is to be used for interpolation near the end of a table, θ must be given a negative value; if it is used for *extrapolating* beyond the end of a table, θ must be given a positive value.

Difference operators

The operators used here obey the rules of algebra provided that the functions they operate on are polynomials. We can, therefore, avail ourselves of known or provable results from algebra and trigonometry to deduce formulae in finite differences. These formulae are established only when the functions concerned are polynomials, although they are intended to be used for purposes of approximation with functions which can be represented by polynomials to the accuracy required over a limited range.

We introduce first the operators D and E:

$$Df_j = f_j' \quad (D = d/dx)$$
$$Ef_j = f_{j+1}$$

Thus E stands for the operation of striking out the suffix j and replacing it by $j+1$.

We have
$$DE = ED$$
as may be verified as follows

$$DEf_0 = Df_1 = f_1'$$
$$EDf_0 = Ef_0' = f_1'$$

Now
$$f_1 = f(x_0+h)$$
$$= f(x_0)+hf'(x_0)+\frac{h^2}{2!}f''(x_0)+\ldots$$

(this series terminates since $f(x)$ is a polynomial)

$$= f_0+hDf_0+\frac{h^2}{2!}D^2f_0+\ldots$$
$$= e^{hD}f_0$$

(this means no more than that if you follow the rules of algebra and expand e^{hD} you get the right answer)

that is
$$Ef_0 = e^{hD}f_0$$

Thus, we have $E = e^{hD}$ as a relation between operators.

Again
$$\Delta f_0 = f_1-f_0$$
$$= (E-1)f_0$$

so that
$$\Delta = E-1$$

similarly
$$\nabla = 1-E^{-1}$$

Newton–Gregory interpolation formula.

$$\begin{aligned}f(x_0+\theta h) &= f_0+\theta h D f_0+\frac{1}{2!}\theta^2 h^2 D^2 f_0+\ldots\\ &= e^{\theta h D} f_0\\ &= E^{\theta} f_0\\ &= (1+\Delta)^{\theta} f_0\\ &= f_0+\theta\Delta f_0+\frac{1}{2!}\theta(\theta-1)\Delta^2 f_0+\ldots\end{aligned}$$

This is the formula that was obtained earlier by an elementary method.

If, instead of putting $E = 1+\Delta$ we put $E = (1-\nabla)^{-1}$ we obtain on expansion the backward difference formula

$$f(x_0+\theta h) = f_0+\theta\nabla f_0+\frac{1}{2!}\theta(\theta+1)\nabla^2 f_0+\ldots$$

More difference operators. Define $E^{\frac{1}{2}}$ as the operation of increasing a suffix by $\frac{1}{2}$; then

$$E^{\frac{1}{2}}E^{\frac{1}{2}} = E$$

Similarly, $E^{-\frac{1}{2}}$ is defined as the operation of decreasing a suffix by $\frac{1}{2}$. Next define

$$\delta = E^{\frac{1}{2}}-E^{-\frac{1}{2}}$$

then

$$\begin{aligned}\delta f_{\frac{1}{2}} &= E^{\frac{1}{2}} f_{\frac{1}{2}}-E^{-\frac{1}{2}} f_{\frac{1}{2}}\\ &= f_1-f_0\end{aligned}$$

Thus δ is the same as the δ we have already used in defining central differences.

Note that

$$\begin{aligned}\delta^2 &= (E^{\frac{1}{2}}-E^{-\frac{1}{2}})(E^{\frac{1}{2}}-E^{-\frac{1}{2}})\\ &= E-2+E^{-1}\end{aligned}$$

so that

$$\delta^2 f_0 = f_1-2f_0+f_{-1}$$

With a qualification to be made later, there is no occasion to use odd powers of δ operating on function values with integral suffixes, or even powers of δ operating on function values with half-integral suffixes.

A relation between D and δ.

$$\begin{aligned}\delta &= E^{\frac{1}{2}}-E^{-\frac{1}{2}}\\ &= e^{\frac{1}{2}hD}-e^{-\frac{1}{2}hD} \quad \text{since} \quad E = e^{hD}\\ &= 2\sinh(\tfrac{1}{2}hD)\end{aligned}$$

Thus $$hD = 2\sinh^{-1}(\tfrac{1}{2}\delta),$$

or $$hD = \delta(1-\tfrac{1}{24}\delta^2+\tfrac{3}{640}\delta^4-\ldots)$$

Everett's formula

This is an interpolation formula based on central differences. It is one of many possible formulae, difference notation being highly redundant. Everett's formula is distinguished by having only even-order differences, and we therefore start with an expression of the form

$$f_\theta = \bar{\theta}f_0+\theta f_1+E_2\delta^2 f_0+F_2\delta^2 f_1+E_4\delta^4 f_0+F_4\delta^4 f_1+\ldots \qquad (1)$$

where E_2, F_2, E_4, F_4, ... are functions of θ to be determined.

The following relations between operators can readily be established.

$$\delta^2 = E^{-1}\Delta^2; \quad \delta^2 = \frac{\Delta^2}{1+\Delta}$$

Hence $$\delta^2 f_0 = \frac{\Delta^2}{1+\Delta}f_0$$

$$\delta^2 f_1 = \Delta^2 f_0$$

$$\delta^4 f_0 = \frac{\Delta^4}{(1+\Delta)^2}f_0$$

$$\delta^4 f_1 = \frac{\Delta^4}{1+\Delta}f_0$$

Substituting these values in (1), we obtain

$$f_\theta = \bar{\theta}f_0+\theta f_1+E_2\frac{\Delta^2}{1+\Delta}f_0+F_2\Delta^2 f_0$$
$$+E_4\frac{\Delta^4}{(1+\Delta)^2}f_0+F_4\frac{\Delta^4}{1+\Delta}f_0+\ldots$$

This must be equivalent to $$f_\theta = (1+\Delta)^\theta f_0$$

The first two terms of the two formulae are easily seen to be equivalent. We multiply both formulae by $1+\Delta$ and equate coefficients of Δ^3. This gives

$$\binom{\theta+1}{3} = F_2$$

i.e. $$F_2 = \frac{1}{3!}\theta(\theta^2-1)$$

Similarly, multiplying by $(1+\Delta)^2$ and equating coefficients of Δ^5 we obtain

$$\binom{\theta+2}{5} = F_4$$

i.e.
$$F_4 = \frac{1}{5!}\theta(\theta^2-1)(\theta^2-4)$$

It follows, from symmetry, that

$$E_2 = \frac{1}{3!}\bar{\theta}(\bar{\theta}^2-1)$$

$$E_4 = \frac{1}{5!}\bar{\theta}(\bar{\theta}^2-1)(\bar{\theta}^2-4)$$

Higher order coefficients may be obtained in the same way.

Example. To find log 4·175 from Table 3.

If $\theta = \frac{1}{2}$, Everett's formula becomes

$$f_{\frac{1}{2}} = \tfrac{1}{2}(f_0+f_1)-\tfrac{1}{16}(\delta^2 f_0+\delta^2 f_1)+\tfrac{3}{256}(\delta^4 f_0+\delta^4 f_1)+\ldots$$

Entering the table we obtain

x		f		δ^2		δ^4
4·15		·61804 81		−630		−3
4·20		·62324 93		−616		2
		↓		↓		↓
$f(4{\cdot}175)$	=	·62064 87	+	·00000 78	+	·00000 00
	=	·62065 65				

Notes. (1) The Everett formula, like the Newton–Gregory formula, is equivalent to linear interpolation plus corrections.

(2) The terms fall off more rapidly than in the Newton–Gregory formula since the coefficients are smaller. This is the way in which the expected advantage of using a formula that is symmetrical about the interval x_0 to x_1 is realized.

(3) Everett's formula truncated at a given order of differences is exactly equivalent to the corresponding Lagrange formula, and fits the same polynomial. But with Everett one simply has to add further terms to go to a higher order, whereas with Lagrange one has to begin again with a wholly different formula.

Indication that we have gone to a high enough order, and that a polynomial has been satisfactorily fitted, is obtained when the addition of further terms ceases to make any difference. If this happens, we stop taking further terms, and we can have confidence in the result. If it does not happen, it is a sign that the interval h is too large for interpolation to the accuracy being tried for.

If we persist in continuing beyond the point at which additional terms have become negligible, we shall eventually find that they become large again. It has, however, already been pointed out that the higher differences depend on the rounding errors, rather than on the function tabulated. By including these terms, therefore, we will simply be spoiling the result. Finite difference series thus behave more like asymptotic series than like convergent series. Nevertheless, one speaks of a finite difference series converging when a point is reached at which the terms become negligible. An error made by not including enough terms is called a truncation error.

Whether a given order of differences can be neglected depends on the magnitudes of the coefficients (and hence on θ) as well as on the magnitudes of the differences. As a rough rule, however, one can certainly neglect terms in Everett's formula if the differences are less than in the following table:

Difference neglected	Limiting value
2nd	4
4th	20
6th	100

If differences had to be computed specially, this would be a point against interpolation formulae using differences. But in published tables, the differences are often given. (In old-fashioned desk work, the computer usually needed differences in any case for checking.)

Everett's formula lends itself particularly well to published tables since it is only necessary to add to the column of function values a column giving values of second differences (and possibly also a column giving fourth differences) in order to make it unnecessary for the user to compute any differences himself. For example, the following table contains sufficient information to enable interpolation to be performed anywhere in the range of tabulation without the computation of differences:

x	Log x	δ^2
5·00	·6989700	−434
5·05	·7032914	−426
5·10	·7075702	−418
5·15	·7118072	−409

Many formulae in numerical analysis can be expressed either in terms of function values, or in terms of differences. The former is called the *Lagrangean* form. Numerically the two forms are exactly equivalent. The difference form gives more insight into the action of a formula, but the Lagrangean form is more efficient when it comes to actual programming. There is, in fact, hardly ever any point in forming a difference table explicitly in the store of a digital computer. Thus one thinks and speaks in terms of differences, but writes one's programs in terms of function values.

Bessel's formula

This formula, which is worth mentioning although it is not now of great practical importance, is as follows:

$$f_\theta = \tfrac{1}{2}(f_0+f_1)+(\theta-\tfrac{1}{2})\delta f_{\frac{1}{2}}+B_2(\delta^2 f_0+\delta^2 f_1)$$
$$+B_3\delta^3 f_{\frac{1}{2}}+B_4(\delta^4 f_0+\delta^4 f_1)+\dots$$

where

$$B_2 = \frac{1}{2.2!}\theta(\theta-1)$$

$$B_3 = \frac{1}{3!}\theta(\theta-\tfrac{1}{2})(\theta-1)$$

$$B_4 = \frac{1}{2.4!}(\theta+1)\theta(\theta-1)(\theta-2)$$

This is equivalent to Everett's formula as may be seen by putting

$$\delta^3 f_{\frac{1}{2}} = \delta^2 f_1 - \delta^2 f_0 \quad \text{etc.}$$

As before the first two terms correspond to linear interpolation. Four terms of Everett are equivalent to four terms of Bessel, i.e. they allow for third differences.

It is convenient to define an averaging operation by the operator

$$\mu = \tfrac{1}{2}(E^{-\frac{1}{2}}+E^{\frac{1}{2}}) = \cosh(\tfrac{1}{2}hD)$$

We then have $\delta^2 f_0+\delta^2 f_1 = 2\mu\delta^2 f_{\frac{1}{2}}$, etc., and Bessel's formula becomes

$$f_\theta = \mu f_{\frac{1}{2}}+(\theta-\tfrac{1}{2})\,\delta f_{\frac{1}{2}}+2B_2\delta^2\mu f_{\frac{1}{2}}+B_3\delta^3 f_{\frac{1}{2}}+2B_4\delta^4\mu f_{\frac{1}{2}}+\dots$$

The use of μ makes necessary the qualification referred to earlier of the rule that even powers of δ are only used to operate on qualities with integral suffixes, and odd powers of δ only on qualities with half integral suffixes.

Throwback

This is a device, introduced into practical computing by L. J. Comrie, whereby fourth differences which are not too large can be allowed for by using modified values of the corresponding second differences.

Consider Everett's formula as far as fourth differences:

$$f_\theta = \bar{\theta}f_0 + \theta f_1 + E_2\delta^2 f_0 + F_2\delta^2 f_1 + E_4\delta^4 f_0 + F_4\delta^4 f_1 + \ldots$$

The terms in $\delta^2 f_0$ and $\delta^4 f_0$ may be grouped together as follows:

$$E_2[\delta^2 f_0 + (E_4/E_2)\delta^4 f_0] = E_2[\delta^2 f_0 + \tfrac{1}{20}(\bar{\theta}^2 - 4)\delta^4 f_0]$$

Over the range $0 \leqslant \bar{\theta} \leqslant 1$ the quantity $\frac{1}{20}(\bar{\theta}^2 - 4)$ is small and does not change greatly, varying in fact between $-{\cdot}2$ and $-{\cdot}15$. Quite a good approximation may, therefore, be obtained by giving it a constant value somewhere between these limits. The value taken in practice is $-{\cdot}184$ which may be shown to minimize the maximum error. Accordingly we write $\delta_m^2 = \delta^2 - {\cdot}184\delta^4$ and obtain the following approximate form of Everett's formula

$$f_\theta = \bar{\theta}f_0 + \theta f_1 + E_2\delta_m^2 f_0 + F_2\delta_m^2 f_1$$

Note that this formula differs from Everett's formula taken as far as second difference only in that *modified* instead of true differences are used. In a printed table, therefore, modified differences may be given instead of true differences, without the user needing to pay any regard to this fact, as far as the use of Everett's formula is concerned. Throwback allows fourth differences up to about 1000 units to be taken into account; this may be compared with the limiting value of about 20 below which fourth differences may be neglected altogether. Throwback may also be used with Bessel's formula.

Table 6 is a table of sines tabulated at interval 20° in the argument with a column of modified second differences. If used with Everett's formula this table is equivalent to an ordinary four-figure table of sines, and it illustrates the very considerable condensation that can be obtained if one is prepared to use higher order interpolation. Values of the Everett interpolation coefficient at interval ·1 are given in Table 5. The following example illustrates the use of Everett's formula.

To find sin 24°.

$\theta = 4/20 = {\cdot}2$ so that, from Table 5,

$$E_2 = -{\cdot}0480 \quad F_2 = -{\cdot}0320$$

and from Table 6

$$\sin 20^\circ = \cdot 3420; \quad \delta_m^2 = -\cdot 0421$$

$$\sin 40^\circ = \cdot 6428; \quad \delta_m^2 = -\cdot 0794$$

Everett's formula gives

$$\begin{aligned}\sin 24^\circ &= (1-\cdot 2)\times\cdot 3420+\cdot 2\times\cdot 6428\\ &\quad +(-\cdot 0480)\times(-\cdot 0421)+(-\cdot 0320)\times(-\cdot 0794)\\ &= \cdot 4067\end{aligned}$$

TABLE 5

θ	E_2	F_2
·1	−·0285	−·0165
·2	−·0480	−·0320
·3	−·0595	−·0455
·4	−·0640	−·0560
·5	−·0625	−·0625
·6	−·0560	−·0640
·7	−·0455	−·0595
·8	−·0320	−·0480
·9	−·0165	−·0285

TABLE 6

x	Sin x	δ_m^2
0°	·0000	0
20°	·3420	−421
40°	·6428	−794
60°	·8660	−1067
80°	·9848	−1214
100°	·9848	−1214

Neville's process. This is a method of performing interpolation in such a way that it is possible to pass successively to higher orders of interpolation by repetition of what is essentially the same process; neither interpolation coefficients nor differences appear explicitly. Let $f_{0,1}$ be the first approximation to the value of f at the point $x_0+\theta h$; then

$$f_{0,1} = \begin{vmatrix} \theta & f_0 \\ \theta-1 & f_1 \end{vmatrix}$$

This clearly corresponds to linear interpolation between the values f_0 and f_1. Similarly, we define

$$f_{1,2} = \begin{vmatrix} \theta-1 & f_1 \\ \theta-2 & f_2 \end{vmatrix}$$

This corresponds to a linear extrapolation based on the values f_1 and f_2. The next approximation to the interpolated value is $f_{0,1,2}$ defined by

$$f_{0,1,2} = \tfrac{1}{2}\begin{vmatrix} \theta & f_{0,1} \\ \theta-2 & f_{1,2} \end{vmatrix}$$

It is not difficult to see that the right-hand side is a quadratic in θ that takes the values f_0, f_1, f_2 when $\theta = 0, 1, 2$. $f_{0,1,2}$ therefore corresponds to quadratic interpolation. Similarly, we define

$$f_{-1,0,1} = \tfrac{1}{2}\begin{vmatrix} \theta+1 & f_{-1,0} \\ \theta-1 & f_{0,1} \end{vmatrix}$$

The next approximation, corresponding to cubic interpolation, is

$$f_{-1,0,1,2} = \tfrac{1}{3}\begin{vmatrix} \theta+1 & f_{-1,0,1} \\ \theta-2 & f_{0,1,2} \end{vmatrix}$$

The process may be continued as far as desired; normally it would be terminated when two successive values for the interpolate were in agreement.

Examples

1. Find the values of log 4·623 and log 5·108 by interpolating into Table 3 using (*a*) Everett, (*b*) Lagrange, and (*c*) Neville. [·6649239, ·7082509]

2. Find log 4·025 by interpolating into Table 3 using the Newton–Gregory formula. [·6047658]

3. The following table of solutions of $y^3 = y+a$ contains an error. Locate it, and estimate a correction by means of differences. Verify your estimate by evaluating the correct value, using the Newton–Raphson method, or otherwise solving the original equation for the relevant value of a.

a	y	a	y
6	2·000	18	2·748
8	2·166	20	2·837
10	2·309	22	2·924
12	2·435	24	3·000
14	2·548	26	3·075
16	2·652	28	3·146
		30	3·214

[Cambridge Mathematical Tripos I, 1961]

4. Use difference operators to derive the following interpolation formula

$$f_\theta = f_1 - \bar{\theta}\nabla f_1 + \frac{1}{2!}\bar{\theta}(\bar{\theta}-1)\nabla^2 f_1 - \ldots$$

5. If $f_\theta = \phi_0(\delta)f_0+\phi_1(\delta)f_1$ where ϕ_0 and ϕ_1 are even functions of δ, show that

$$\phi_0 = \sinh \bar{\theta}U/\sinh U \quad \text{and} \quad \phi_1 = \sinh \theta U/\sinh U$$

where $U = 2\sinh^{-1}(\frac{1}{2}\delta)$. Hence derive Everett's formula given the identity

$$\sinh \beta u/\sinh u = \beta\left[1+\frac{1}{3!}(\beta^2-1)z^2+\frac{1}{5!}(\beta^2-1)(\beta^2-4)z^4+\dots\right]$$

where $z = 2\sinh(\frac{1}{2}u)$.

6. Show that if $\delta_m^2 = \delta^2+C\delta^4$ is used in place of δ^2 in Bessel's formula, and if differences higher than the fourth are negligible, then the error introduced by omitting the term in δ^4 is

$$B_2[\tfrac{1}{12}(\theta+1)(\theta-2)-C](\delta^4 f_0+\delta^4 f_1)$$

Show that if $C=-(3+\sqrt{2})/24=-\cdot183925\dots$, the coefficient of $\delta^4 f_0+\delta^4 f_1$ in the expression has three numerically equal positive and negative extreme values in the range $0\leqslant 0\leqslant 1$ and that the maximum absolute error is then a minimum.

7. Define

$$f(x;\, x_0, x_1) = \frac{1}{x_1-x_0}\begin{vmatrix} x-x_0 & f(x_0) \\ x-x_1 & f(x_1) \end{vmatrix}$$

$$f(x;\, x_0, x_1, x_2) = \frac{1}{x_2-x_0}\begin{vmatrix} x-x_0 & f(x;\, x_0, x_1) \\ x-x_2 & f(x;\, x_1, x_2) \end{vmatrix}$$

and generally

$$f(x;\, x_0, x_1, \dots, x_n) = \frac{1}{x_n-x_0}\begin{vmatrix} x-x_0 & f(x;\, x_0, x_1, \dots, x_{n-1}) \\ x-x_n & f(x;\, x_1, x_2, \dots, x_n) \end{vmatrix}$$

Prove, by induction or otherwise, that $f(x;\, x_0, x_1, \dots, x_n)$ is a polynomial of degree not exceeding n and that if x_r stands for one of the values $x_0, x_1, \dots, x_n$ then $f(x_r;\, x_0, x_1, \dots, x_n) = f(x_r)$. Show further, by using the theorem that the polynomial of degree at most n through $n+1$ distinct points is unique, that the value of $f(x;\, x_0, x_1, \dots, x_n)$ depends only on the points and not on the order in which they are labelled or used. (This is a more general statement of Neville's process than that given in the text; in particular the points x_r need not be equally spaced.)

4

NUMERICAL INTEGRATION AND DIFFERENTIATION

Numerical integration or quadrature

If a curve can be replaced with sufficient precision by a polynomial, the area under the curve can be found by integrating the polynomial. Normally, we work with a curve defined by a number of isolated points as shown. This may be the way the function is given, or we may ourselves have computed the points from a formula; in the latter case we clearly have full control over the number of points and their spacing.

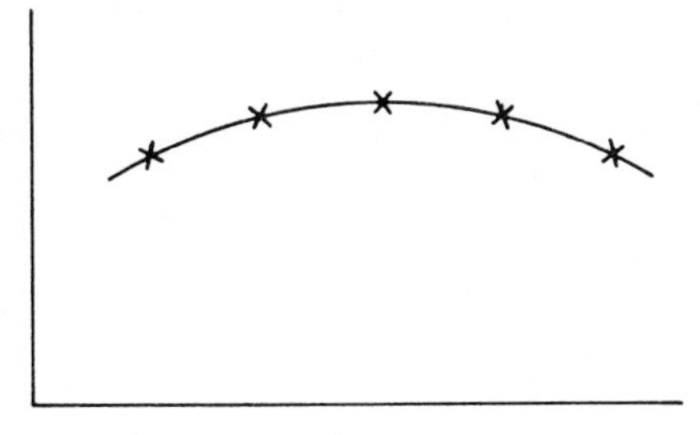

Fig. 1

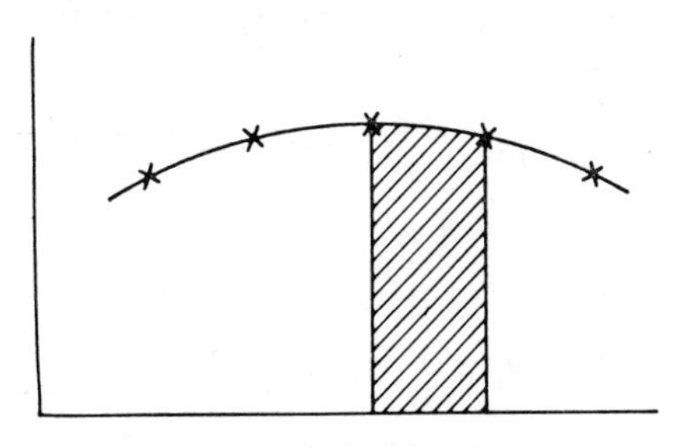

Fig. 2

We shall first concern ourselves with a formula giving the area of a single strip as shown. The total area can then be found by adding up the areas of a number of strips. We could proceed by integrating, say, Bessel's formula, but it is just as convenient to start *ab initio*.

Integration over a single strip

We quote the following formula

$$\frac{1}{x}\tanh x = 1-\tfrac{1}{3}x^2+\tfrac{2}{15}x^4-\dots$$

and recollect that

$$hD = 2\sinh^{-1}(\tfrac{1}{2}\delta)$$
$$= \delta(1-\tfrac{1}{24}\delta^2+\tfrac{3}{640}\delta^4-\dots)$$

$$I = \int_{x_0}^{x_1} f(x)\,dx$$
$$= \Big[D^{-1}f\Big]_{x_0}^{x_1}$$
$$= D^{-1}f_1-D^{-1}f_0$$
$$= (E-1)D^{-1}f_0$$
$$= \tfrac{1}{2}h\frac{E-1}{E+1}\frac{1}{\tfrac{1}{2}hD}(E+1)f_0$$

Now

$$\frac{E-1}{E+1}\frac{1}{\tfrac{1}{2}hD} = \frac{e^{hD}-1}{e^{hD}+1}\frac{1}{\tfrac{1}{2}hD}$$
$$= \frac{1}{\tfrac{1}{2}hD}\tanh(\tfrac{1}{2}hD)$$
$$= 1-\frac{1}{3.2^2}(hD)^2+\frac{2}{15.2^4}(hD)^4-\dots$$
$$= 1-\frac{1}{3.2^2}\delta^2(1-\tfrac{1}{24}\delta^2+\tfrac{3}{640}\delta^4-\dots)^2$$
$$+\frac{2}{15.2^4}\delta^4(1-\tfrac{1}{24}\delta^2+\tfrac{3}{640}\delta^4-\dots)^4-\dots$$
$$= 1-\tfrac{1}{12}\delta^2+\tfrac{11}{720}\delta^4-\dots$$

$$\int_{x_0}^{x_1} f\,dx = \tfrac{1}{2}h(1-\tfrac{1}{12}\delta^2+\tfrac{11}{720}\delta^4-\dots)(E+1)f_0$$
$$= \tfrac{1}{2}h[(f_0+f_1)-\tfrac{1}{12}(\delta^2f_0+\delta^2f_1)+\tfrac{11}{720}(\delta^4f_0+\delta^4f_1)-\dots] \quad (1)$$

The first part of the formula, namely $\tfrac{1}{2}h(f_0+f_1)$ corresponds to the crudest of approximations obtained by joining the two points by a straight line so that the strip is a trapezium. It is known as the *trapezoidal rule* (for some reason, quadrature formulae are known as rules). The succeeding terms may be regarded as corrections. Note that in order to compute δ^2, δ^4, etc., it is necessary to take points outside the range x_0 to x_1.

Alternative treatment. We start with the trapezoidal rule

$$I = \int_{x_0}^{x_1} f dx = \tfrac{1}{2}h(f_0+f_1) \tag{2}$$

which is clearly exact for $f = a+bx$

We apply (2) to $f = a+bx+cx^2+dx^3$

and calculate the error, which we afterwards use as a correction. Since (2) is linear we may drop the first two terms and take simply

$$f = cx^2+dx^3$$

It is convenient to choose the origin of x so that $x_0 = 0$.

We then have:

$$f_{-1} = ch^2-dh^3 \quad f_0 = 0 \quad f_1 = ch^2+dh^3$$

$$f_2 = 4ch^2+8dh^3$$

and it is easily shown that

$$\delta^2 f_0 = 2ch^2 \quad \delta^2 f_1 = 2ch^2+6dh^3$$

The true value of I is $\tfrac{1}{3}ch^3+\tfrac{1}{4}dh^4$

(2) gives $I = \tfrac{1}{2}h(ch^2+dh^3)$

The error is $(\tfrac{1}{6}ch^3+\tfrac{1}{4}dh^4)$

$$= \tfrac{1}{24}h(\delta^2 f_0+\delta^2 f_1)$$

Applying this as a correction to (2), we have

$$\int_{x_0}^{x_1} f dx = \tfrac{1}{2}h[(f_0+f_1)-\tfrac{1}{12}(\delta^2 f_0+\delta^2 f_1)]$$

Further terms may be obtained in a similar manner.

Example. Using Table 3, to find $\int_{4\cdot05}^{4\cdot10} \log_{10} x\,dx$

x	Log x	δ^2
4·05	·6074550	−·0000661
4·10	·6127839	−·0000647

$$\int_{4\cdot05}^{4\cdot10} \log_{10} x\,dx = \tfrac{1}{2}\times\cdot05(1\cdot2202389+\cdot0000109)$$

$$= \cdot03050\,6245$$

Since $\int \log_{10} x\,dx = x \log_{10} x - mx$ where $m = \cdot 4342945$

it is possible to evaluate $\int_{4\cdot 05}^{4\cdot 10} \log_{10} x\,dx$ otherwise using the same table. This provides an object lesson in the effects of rounding errors.

$$\Big[x \log_{10} x - mx\Big]_{4\cdot 05}^{4\cdot 10} = 4\cdot 10 \times \cdot 61278\,39 \;①$$
$$- 4\cdot 05 \times \cdot 60745\,50 \;②$$
$$- \cdot 05 \times \cdot 43429\,45$$
$$= \cdot 03050\,65$$

which may be compared with the value obtained above, namely ·03050 6245.

The latter is in fact more accurate, as is shown by a calculation using 10-figure logarithms which gives ·03050 624. The explanation is that calculation of the definite integral using the indefinite integral involves the subtraction of the numbers ① and ② which are of similar magnitude, and some of the leading digits are consequently lost. The rounding errors, on the other hand, can, if they happen to be of opposite sign, reinforce one another. The rounding errors in ① and ② can amount to slightly more than two units, i.e. half a unit multiplied by 4·10 or 4·05, and these are amply sufficient to account for the total error in the result. In doing the numerical integration, no subtraction of numbers of comparable magnitude was involved. This brings out the fact that quadrature is a satisfactory process in which one does not tend to lose accuracy through cancellation of leading digits.

When a large number of independent rounding errors are combined, it is necessary to consider their effect statistically. A single error is uniformly distributed in the range $\pm \cdot 5$ units and has a standard error of $1/\sqrt{12}$. If n such errors are added, the resulting error will be normally distributed in the range $\pm \cdot 5n$ with a standard error of $\sqrt{(n/12)}$. Thus, if 10 000 errors are combined, the result *could* be in error by as much as 5 000 units, but since the standard error is $100/\sqrt{12}$, or about 30, there is a million to one chance against the error exceeding 150, while the chance against it exceeding 300 is 10^{23} to 1. Even the most cautious numerical analyst will sometimes be prepared to put his money on when the odds against losing are as long as this.

Extended range of integration. If the single strip formula just obtained is applied to a number of adjacent strips and the results added, it is found that when use is made of the identities

$$\delta^2 f_r = \delta f_{r+\frac{1}{2}} - \delta f_{r-\frac{1}{2}}$$
$$\delta^4 f_r = \delta^3 f_{r+\frac{1}{2}} - \delta^3 f_{r-\frac{1}{2}}$$

some of the terms cancel. We then have

$$\begin{aligned}\int_{x_0}^{x_n} f\,dx = & \tfrac{1}{2}h(f_0+2f_1+2f_2+\ldots+2f_{n-1}+f_n) \\ & +\frac{h}{24}(\delta f_{-\frac{1}{2}}+\delta f_{\frac{1}{2}}-\delta f_{n-\frac{1}{2}}-\delta f_{n+\frac{1}{2}}) \\ & -\frac{11h}{1440}(\delta^3 f_{-\frac{1}{2}}+\delta^3 f_{\frac{1}{2}}-\delta^3 f_{n-\frac{1}{2}}-\delta^3 f_{n+\frac{1}{2}}) \\ & +\ldots \end{aligned} \qquad (3)$$

Thus, apparently, the corrections to the trapezoidal rule depend only on the odd order differences at the ends of the range. There are, however, hidden dangers in the use of the above formula. It is derived on the assumption that the finite difference series for the individual strips converge. If one or more fail to do so, the result will be in error, even if the series in (3) converges. Thus convergence of (3) does not guarantee the accuracy of the result.

An extreme example is supplied by the integral $\int_0^X e^{-x^2}\,dx$ for X large. The odd differences vanish at $x = 0$ and tend to zero at X. If X is very large, it would thus seem that the corrections to the trapezoidal rule are negligible, whatever the value taken for h. This is obviously false and the rule will break down when h becomes large enough, although for small and moderate values of h it will be found to be much more accurate than in the case of a general integrand. Thus, while (3) may be useful as a way of saving computing time, compared with the application of (1) to each strip separately, an independent check should always be made of its accuracy.

Integration over two strips

We now derive a formula for the area of two adjacent strips. If the formula is to be applied to an extended integral, the total number of strips must be even.

We quote
$$\sinh\theta = \theta+\frac{\theta^3}{3!}+\frac{\theta^5}{5!}+\ldots$$

$$\int_{x_{-1}}^{x_1} f(x)\,dx = (E-E^{-1})\,D^{-1}f_0$$
$$= 2\sinh hD\; D^{-1}f_0$$
$$= 2h\frac{\sinh hD}{hD}f_0$$

$$\frac{\sinh hD}{hD} = 1+\frac{1}{3!}(hD)^2+\frac{1}{5!}(hD)^4+\ldots$$
$$= 1+\frac{\delta^2}{3!}(1-\tfrac{1}{24}\delta^2+\tfrac{3}{640}\delta^4-\ldots)^2$$
$$+\frac{\delta^4}{5!}(1-\tfrac{1}{24}\delta^2+\tfrac{3}{640}\delta^4-\ldots)^4+\ldots$$
$$= 1+\tfrac{1}{6}\delta^2-\tfrac{1}{180}\delta^4+\tfrac{1}{1512}\delta^6-\ldots$$

$$\int_{x_{-1}}^{x_1} f\,dx = 2h(f_0+\tfrac{1}{6}\delta^2 f_0-\tfrac{1}{180}\delta^4 f_0+\tfrac{1}{1512}\delta^6 f_0-\ldots) \tag{1}$$

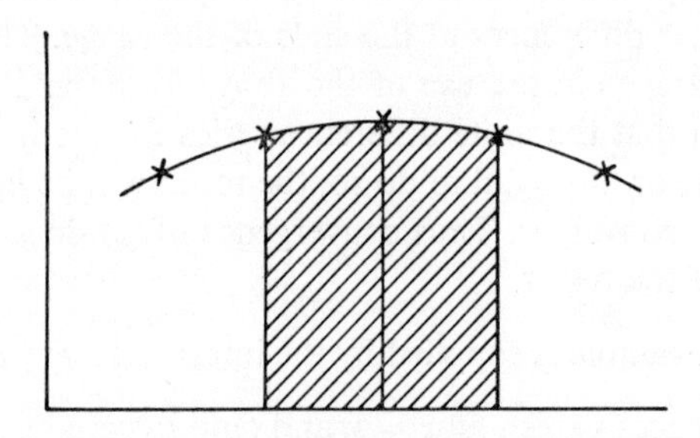

Fig. 3

The first term $2hf_0$ represents an obvious but crude approximation, and succeeding terms may be regarded as corrections to this. Note the rapidity with which the coefficients fall away. In this case, it is possible to include terms as far as that in δ^2 without using points outside the double strip. If we do this, we have

$$\int_{x_{-1}}^{x_1} f\,dx = 2h(f_0+\tfrac{1}{6}\delta^2 f_0)$$
$$= \tfrac{1}{3}h(f_{-1}+4f_0+f_1)$$

which will be recognized as Simpson's rule.

Alternative treatment. We start with the crude approximation

$$I = \int_{x_{-1}}^{x_1} f dx = 2hf_0 \tag{2}$$

which is easily shown to be exact for

$$f = a+bx$$

We apply (2) to $\quad f = a+bx+cx^2+dx^3$

or, omitting linear terms, to

$$f = cx^2+dx^3$$

and calculate the error. As before, it is convenient to choose the origin of x so that $x_0 = 0$ and it is easily seen that $\delta^2 f_0 = 2ch^2$.

The true value of I is $\frac{2}{3}ch^3$ whereas (2) gives $I = 0$.

The error is $-\frac{2}{3}ch^3$ or $-\frac{1}{3}h\delta^2 f_0$. Applying this as a correction to (2), we have

$$I = 2h(f_0+\tfrac{1}{6}\delta^2 f_0)$$

Further terms may be similarly obtained.

Accuracy. Since (1) contains even differences only, the formula obtained by truncating it at any point includes the effect of the next (odd) difference. Thus, Simpson's rule, for example, is exact for a polynomial of degree less than or equal to three, not merely two.

It is possible to derive formulae (known as Newton–Cotes formulae) for integration over a larger number of equal strips although, for reasons which will be mentioned later, their use is not generally to be recommended.

People sometimes talk of one formula being more 'accurate' than another; for example, of Simpson's rule being more 'accurate' than the trapezoidal rule. This, however, can be misleading, since both the trapezoidal rule and Simpson's rule can be made to give any desired number of correct decimals if h is chosen appropriately. What is true is that a smaller value of h (and more strips) may be needed in the case of one formula than are needed in the case of another.

Gauss's formula

Integration formulae can be expressed in terms of function values as was done above for Simpson's rule. In general, we may take

$$\int_a^b f(x)\,dx = H_0 f(x_0)+H_1 f(x_1)+\ldots+H_n f(x_n) \tag{1}$$

where $n+1$ ordinates are used, and if these are taken at specified points (for example, if they are equally spaced as has been assumed in the work so far) the best that can be done, in general, is to choose the $(n+1)$ coefficients $H_0, H_1, \ldots, H_n$ so as to make (1) exact when $f(x)$ is an arbitrary polynomial of degree not greater than n. It will now be shown by an indirect method that if $x_0, x_1, \ldots, x_n$ are also regarded as adjustable, then (1) may be made exact for all polynomials of degree not greater than $2n+1$.

It is convenient to take the range of integration to be -1 to $+1$. We then have

$$\int_{-1}^{1} f(x)\,dx = H_0 f(x_0) + H_1 f(x_1) + \ldots + H_n f(x_n) \tag{2}$$

If $x_0, x_1, \ldots, x_n$ are *given* the values of $H_0, H_1, \ldots, H_n$ which make (2) exact for a polynomial of degree less than or equal to n may be determined as follows.

Let
$$\pi(x) = (x-x_0)(x-x_1)\ldots(x-x_n)$$

Take
$$f(x) = \frac{\pi(x)}{x-x_r}$$

and substitute in (2). Then

$$\int_{-1}^{1} \frac{\pi(x)}{x-x_r}\,dx = H_r \operatorname*{Lt}_{x\to x_r} \frac{\pi(x)}{x-x_r} = H_r \pi'(x_r)$$

i.e.
$$H_r = \frac{1}{\pi'(x_r)} \int_{-1}^{1} \frac{\pi(x)}{x-x_r}\,dx \tag{3}$$

With a given set of x_r and these values of H_r, (2) is exact for any polynomial of degree less than or equal to n.

It will now be shown that if $x_0, x_1, \ldots, x_n$ are chosen to be the roots of the Legendre polynomial of degree $n+1$, i.e. if we take

$$\pi(x) = kP_{n+1}(x)$$

where k is constant, then (2) is exact for any polynomial of degree less than or equal to $2n+1$.

Recall that
$$\int_{-1}^{1} \phi(x) P_{n+1}(x)\,dx = 0 \tag{4}$$

if $\phi(x)$ is a polynomial of degree less than $n+1$.

With the H_r chosen according to (3), (2) will be exact for all polynomials of degree less than or equal to n; i.e.

$$\int_{-1}^{1} g(x)\,dx = \sum_{0}^{n} H_r g(x_r) \quad (\text{degree of } g \leqslant n) \tag{5}$$

What we must now show is that, if the x_r are chosen so that $P_{n+1}(x_r) = 0$, (2) is in fact exact for polynomials of degree up to $2n+1$. For this purpose we take for $f(x)$

$$\begin{array}{ccccc} f(x) = g(x) & + & \phi(x) & & P_{n+1}(x) \\ \downarrow & & \downarrow & & \downarrow \\ \text{degree} \leqslant n & & \text{degree} \leqslant n & & \text{degree } n+1 \end{array}$$

where g and ϕ are arbitrary polynomials of the degree indicated. $f(x)$ is a general polynomial of degree $2n+1$. Since $P_{n+1}(x_r) = 0$ it follows that

$$f(x_r) = g(x_r) \tag{6}$$

Now,

$$\begin{aligned} \int_{-1}^{1} f(x)\,dx &= \int_{-1}^{1} g(x)\,dx + \int_{-1}^{1} \phi(x) P_{n+1}(x)\,dx \\ &= \int_{-1}^{1} g(x)\,dx \quad \text{by (4)} \\ &= \sum_0^n H_r g(x_r) \quad \text{by (5) since degree of } g(x) \leqslant n \\ &= \sum_0^n H_r f(x_r) \quad \text{by (6} \end{aligned}$$

This establishes the result.

We have for H_r

$$H_r = \frac{1}{P'_{n+1}(x_r)} \int_{-1}^{1} \frac{P_{n+1}(x)}{x-x_r}\,dx$$

All the H_r are positive as may be readily shown by applying the integration formula (2) to the following polynomial which is of degree $2n+1$

$$\left[\frac{P_{n+1}(x)}{(x-x_s)P'_{n+1}(x_s)}\right]^2$$

This polynomial vanishes for all the x_r, except x_s where it takes the value one: elsewhere it takes only positive values. Thus all the coefficients $H_0, H_1, \ldots, H_n$ are positive. Furthermore, by a property of the Legendre polynomials, all the x_r lie in the range $-1 < x_r < 1$, the end points being excluded. Examples of values of $x_0, x_1, \ldots, x_n$ and $H_0, H_1, \ldots, H_n$ are given below.

Gauss 5-point formula:

$$\begin{array}{ll} x_0 = -x_4 = -{\cdot}9062 & H_0 = H_4 = {\cdot}2369 \\ x_1 = -x_3 = -{\cdot}5385 & H_1 = H_3 = {\cdot}4786 \\ x_2 = 0 & H_2 = {\cdot}5690 \end{array}$$

Gauss 6-point formula:

$$x_0 = -x_5 = -\cdot 9325 \qquad H_0 = H_5 = \cdot 1713$$
$$x_1 = -x_4 = -\cdot 6612 \qquad H_1 = H_4 = \cdot 3608$$
$$x_2 = -x_3 = -\cdot 2386 \qquad H_2 = H_3 = \cdot 4679$$

Convergence of high-order quadrature formulae

A formula such as

$$\int_{-1}^{1} f(x)\,dx = \sum_{0}^{n} H_r f(x_r) \tag{1}$$

in which $-1 \leqslant x_r \leqslant 1$ and $H_r > 0$ may be regarded as giving a weighted mean of the function at the points x_r. For this reason, the H_r are often referred to as the *weights*. Averaging is a highly satisfactory operation as regards accumulation of rounding errors, and the value given for the integral may be more accurate than the individual ordinates.

If a series of approximations to an integral over a given range is computed by repeated use of a formula of the type (1), with an increasing number of points, then the above remark about the weighted mean suggests that, if the weights are all positive, the approximations obtained will get closer and closer to the true value. This in fact may be shown to be the case.

If some of the weights are negative, the above intuitive approach fails, and so does the proof. Newton–Cotes formulae with 8, 10 or more strips have negative weights, and it is easy to give examples for which successive values obtained from formulae of increasing order diverge. If the integrand is given explicitly, a danger sign for divergence is the existence of a singularity just off the real axis.

Consider, for example, the integral $\frac{1}{2}\int_{-4}^{4} \frac{dx}{1+x^2}$ in which the integrand has singularities at $\pm i$. Successive values obtained using Newton–Cotes formulae with 4, 6, 8 and 10 strips are 1·139, 1·664, ·971 and 1·798, which are seen to oscillate with increasing amplitude about the true value which is 1·326.

For the reasons just indicated, it is probably better to avoid the use of Newton–Cotes formulae with more than about 4 strips, and instead to cover a long range of integration in a piece-wise fashion.

Practical checks on accuracy. If an extended range of integration has been divided into a number of subranges to each of which a Newton–

Cotes formula—for example, a single strip or a double strip formula—has been applied, it is necessary to verify that

(1) the finite difference formulae converge in each subrange,

(2) the accumulated error is negligible when the individual contributions are added,

(3) no mistakes or machine errors have occurred.

Two possible methods of checking points (1) and (2) present themselves:

(1) to verify by examination of the neglected terms in the finite difference formula that the result for each step is accurate enough to ensure the accuracy of the sum,

(2) to repeat the entire calculation right to the end with a smaller value of h.

There is no doubt that, when working with a digital computer, the last method is the one to adopt, even though the check involves as much—in fact slightly more—computing time than did the primary calculation. Since the check is made at the very end, the whole of the calculation is embraced by it, and, in particular, a check against machine errors is included. It should, however, be remembered that the check consists in comparing two numbers only, and that the chance that these two numbers may agree while both being in error is not necessarily negligible. If an isolated integral is being computed, it may be thought worth while to perform the calculation for a third value of h.

If the whole range of integration is covered by a single Gauss formula, which may be of high order, the only way of obtaining a check is by repeating the calculation, using, for example, another Gauss formula of order one higher.

For a long range of integration, rather than use a Gauss formula of very high order, it may be more convenient to apply a Gauss formula of modest order (e.g. six) to a series of subintervals, not necessarily equal. A check can then be obtained by repeating the calculation with a different number of subintervals.

When a series of integrals is being computed some economy in checking may be obtained by going in a direction in which the accuracy of the formula used increases. For example, when tabulating the function

$$\mathrm{Ch}(x, \chi) = x \sin \chi \int_0^{\chi} \exp\,(x - x \sin \chi/\sin \lambda)\,\mathrm{cosec}^2\,\lambda\, d\lambda$$

for a fixed value of x, it was found that the accuracy of a formula with given step h increased as χ increased. The program was, therefore, arranged so that for every fifth value of χ, the integration was repeated with $2h$ instead of h. If the result obtained agreed with the previous one, h was again doubled and another value for the integral computed. This procedure was repeated until divergence occurred, when h was halved and this value used for the succeeding four values. An example of results obtained is given below. The figures in brackets are the values of h, expressed in terms of the starting value as 1, with which the preceding function values were computed.

$x = 250$

χ	Ch (x, χ)			
20°	1·0636 (1)	1·0636 (2)	1·0636 (4)	1·0633 (8)
21	1·0705 (4)			
22	1·0778 (4)			
23	1·0856 (4)			
24	1·0938 (4)			
25°	1·1024 (4)	1·1024 (8)	1·0968 (16)	
26	1·1115 (8)			
27	1·1211 (8)			
28	1·1313 (8)			
29	1·1420 (8)			
30°	1·1532 (8)	1·1516 (16)		
31	1·1650 (8)			
...	....			

This method of adjusting the step automatically provides printed evidence on which the accuracy of the final answers can be assessed. It also keeps the interval somewhere near the optimum. This last point is important, since much machine time can be wasted if the interval used is consistently many times smaller than is necessary.

Numerical differentiation

This is a far less satisfactory process than numerical integration, as the following considerations show. If possible it should be avoided altogether. Given two ordinates, as in Fig. 4, the obvious approximation to the derivative is

$$\frac{df}{dx} = \frac{f_1 - f_0}{x_1 - x_0}$$

or

$$f'_{\frac{1}{2}} = \frac{1}{h}\delta f_{\frac{1}{2}} \tag{1}$$

If h is taken to be too small, f_1 and f_0 will be nearly equal and there will not be enough figures in $\delta f_{\frac{1}{2}}$ to give good accuracy; on the other hand, if h is taken to be too large, the chord will not be a good approximation to the tangent.

An improvement can be obtained by recognizing that (1) is the first term of a finite difference series, namely,

$$Df_{\frac{1}{2}} = \frac{1}{h}(\delta - \tfrac{1}{24}\delta^3 + \tfrac{3}{640}\delta^5 - \ldots)f_{\frac{1}{2}} \tag{2}$$

This series enables a larger value of h to be used—if h is too large the series does not converge—but the fundamental difficulty remains. It is not possible to give any simple rule for the choice of h, except to say that it should be as large as is consistent with satisfactory convergence. It may be noted that since second differences are absent from (2), equation (1) is exact for a parabola.

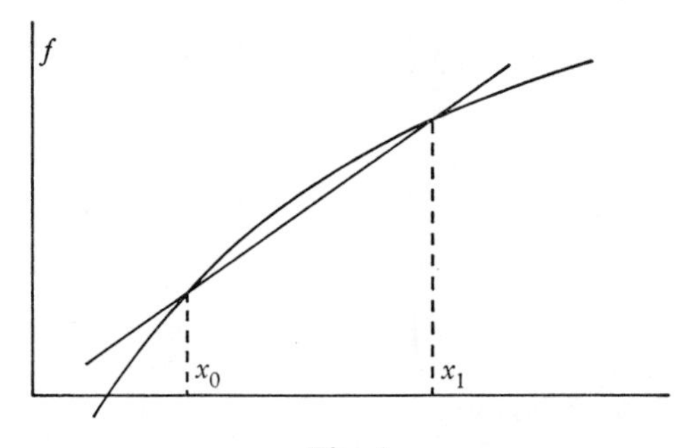

Fig. 4

First derivatives are most naturally computed at the mid-point of the intervals in x, as (1) shows. Derivatives at tabular point can be obtained, if needed, from the following not very rapidly convergent formula

$$f_0' = \frac{1}{h}(\mu\delta f_0 - \tfrac{1}{6}\mu\delta^3 f_0 + \tfrac{1}{30}\mu\delta^5 f_0 - \ldots)$$

which is obtained by expanding

$$Df_0 = \frac{D\mu f_0}{\cosh(\frac{1}{2}hD)} = \frac{2}{h}\frac{\sinh^{-1}\frac{1}{2}\delta}{(1+\frac{1}{4}\delta^2)^{\frac{1}{2}}}\mu f_0$$

Second derivatives, on the other hand, are most naturally computed at tabular values, and the following formula, of quite satisfactory convergence, is available.

$$f_0'' = \frac{1}{h^2}(\delta^2 f_0 - \tfrac{1}{12}\delta^4 f_0 + \tfrac{1}{90}\delta^6 f_0 - \ldots)$$

This is easily obtained by expanding $\frac{1}{h^2}(hD)^2 f_0$. There is an advantage in organizing computations so that, as far as possible, odd order derivatives are avoided.

Examples

1. Obtain the first two terms of the following formula for numerical integration:

$$\int_{x_{-1}}^{x_1} f(x)\,dx = 2h(f_0+\tfrac{1}{6}\delta^2 f_0-\tfrac{1}{180}\delta^4 f_0+\ldots)$$

The following is a table of natural logarithms:

x	Log x
7·0	1·946
8·0	2·079
9·0	2·197
10·0	2·303
11·0	2·398

Use the table to obtain values of

$$\int_8^{10} \log x\,dx$$

(i) by use of the above formula,
(ii) by first integrating $\log x$ in finite terms.
State which of the two values you would expect to be the more accurate, and explain how the discrepancy between the two may be accounted for.
[Cambridge Mathematical Tripos I, 1963]

2. A student obtains a value for $\frac{1}{4}\pi$ by evaluating $\int_0^1 dx/(1+x^2)$ by means of a double application of Simpson's rule, taking ordinates at $x = 0$, ·25, ·5, ·75, 1, and is impressed with the accuracy achieved. Show, by examining the effect of neglected differences, that this accuracy is partly fortuitous. (See Example **4** for a table of $1/(1+x^2)$.)

3. Obtain the following formula for integration over $2k$ strips

$$\int_{x_{-k}}^{x_k} f(x)\,dx = 2hk[1+\tfrac{1}{6}k^2\delta^2+\tfrac{1}{360}k^2(3k^2-5)\delta^4+\ldots]f_0$$

4. Obtain the first two terms of the following formula for the second derivative of a function:

$$f_0'' = \frac{1}{h^2}(\delta^2 f_0-\tfrac{1}{12}\delta^4 f_0+\tfrac{1}{90}\delta^6 f_0-\ldots)$$

The following tables give values of the function $f(x) = (1+x^2)^{-1}$, and their differences, for the intervals h = ·25, h = ·1 and h = ·02, respectively:

x	$f(x)$						
−·25	·94118						
		+5882					
·00	1·00000		−11764				
		−5882		3528			
·25	·94118		−8236		+2826		
		−14118		6354		−5298	
·50	·80000		−1882		−2472		4912
		−16000		3882		−386	
·75	·64000		+2000		−2858		
		−14000		1024			
1·00	·50000		+3024				
		−10976					
1·25	·39024						

x	$f(x)$						
·2	·96154						
		−4411					
·3	·91743		−1125				
		−5536		454			
·4	·86207		−671		−47		
		−6207		407		−40	
·5	·80000		−264		−87		28
		−6471		320		−12	
·6	·73529		+56		−99		
		−6415		221			
·7	·67114		+277				
		−6138					
·8	·60976						

x	$f(x)$				
·44	·83780				
		−1245			
·46	·82535		−16		
		−1261		+3	
·48	·81274		−13		−1
		−1274		+2	
·50	·80000		−11		+2
		−1285		+4	
·52	·78715		−7		−1
		−1292		+3	
·54	·77423		−4		
		−1296			
·56	·76127				

Use the above formula to obtain a value for $f''(\cdot 5)$ from each table.

Compare the values so obtained with the true value, and comment on the accuracy achieved with each of the intervals used.

[Cambridge Mathematical Tripos I, 1964]

5

NUMERICAL SOLUTION OF ORDINARY DIFFERENTIAL EQUATIONS

Initially discussion will be confined to the following first-order differential equation, which is not necessarily linear

$$y' = f(x, y)$$

Predictor-corrector (Adams–Bashforth) method

It is assumed that the integration has already progressed some way, and that a table exists giving values of $y_0, y_{-1}, y_{-2}, \ldots$ and of the corresponding derivatives $f_0, f_{-1}, f_{-2}, \ldots$. The problem is now to compute y_1, and for this purpose two formulae are available:

(1) a *predictor* which gives an estimate for y_1 in terms of values of y already known,

(2) a *corrector* which gives an improved value for y_1 in terms of values of y already known, together with the estimated value given by the predictor.

The Newton–Gregory backward difference formula applied to f is:

$$f_\theta = \left(1+\theta\nabla+\frac{1}{2!}\theta(\theta+1)\nabla^2+\ldots\right)f_0$$

Integrating the left-hand side with respect to θ we have

$$\int f d\theta = (1/h)\int f(x, y)\,dx = y/h$$

It follows, if the whole equation is integrated, that

$$[y] = h[\theta+\tfrac{1}{2}\theta^2\nabla+\tfrac{1}{2}(\tfrac{1}{3}\theta^3+\tfrac{1}{2}\theta^2)\,\nabla^2+\ldots]f_0$$

the limits of integration being left unspecified.

Predictor. This is obtained at once by taking the limits of integration to be $\theta = 0$ and $\theta = 1$:

$$y_1-y_0 = h(1+\tfrac{1}{2}\nabla+\tfrac{5}{12}\nabla^2+\tfrac{3}{8}\nabla^3+\tfrac{251}{720}\nabla^4+\ldots)f_0$$

Corrector. If the limits of integration are taken to be $\theta = -1$ and $\theta = 0$, we have

$$y_0 - y_{-1} = h(1 - \tfrac{1}{2}\nabla - \tfrac{1}{12}\nabla^2 - \tfrac{1}{24}\nabla^3 - \tfrac{19}{720}\nabla^4 - \ldots)f_0$$

The corrector is then obtained by increasing each suffix by 1 and is

$$y_1 - y_0 = h(1 - \tfrac{1}{2}\nabla - \tfrac{1}{12}\nabla^2 - \tfrac{1}{24}\nabla^3 - \tfrac{19}{720}\nabla^4 - \ldots)f_1$$

The following example shows the method applied to the solution of the equation $y' = 2xy + 1$, given $y = 0$ at $x = 0$ and taking $h = \cdot 1$; it is assumed that the integration has been carried as far as $x = \cdot 3$, i.e. that the figures in parentheses have not yet been added.

x	y	y'				
0	0	1				
			201			
·1	·1007	1·0201		420		
			621		49	
·2	·2054	1·0822		469		(42)
			1090		(91)	
·3	·3186	1·1912		(560)		
			(1650)			
·4	(·4453)	(1·3562)				

Application of the predictor then gives

$$\begin{aligned} y_{x=\cdot 4} &= \cdot 3186 + \cdot 1[1\cdot 1912 + \tfrac{1}{2}(\cdot 1090) + \tfrac{5}{12}(\cdot 0469) + \tfrac{3}{8}(\cdot 0049)] \\ &= \cdot 4453 \end{aligned}$$

This value is then entered in the table, the corresponding value of y' computed, and the differences filled in. The corrector is then used to obtain

$$\begin{aligned} y_{x=\cdot 4} &= \cdot 3186 + \cdot 1[1\cdot 3562 - \tfrac{1}{2}(\cdot 1650) - \tfrac{1}{12}(\cdot 0560) - \tfrac{1}{24}(\cdot 0091)] \\ &= \cdot 4455 \end{aligned}$$

In this case the corrector in effect confirms the value given by the predictor, making only a very small correction. The superior convergence of the corrector will be noticed.

The method may be applied using the predictor only, with a sufficiently small step to make the truncation error negligible. The advantages of using the corrector also are:

(1) A slightly larger interval may be used.

(2) A confirmation is obtained that the truncation error is negligible.

(3) Rounding errors are better controlled.

Advantage (2) implies that the correction is quite small, thus limiting advantage (1). The method can be used under conditions in which the corrections are not small, the corrector being used iteratively until convergence is obtained. This, however, is not to be recommended for the following reasons:

(1) Convergence is slow (first order) and repeated use of the corrector does not represent a good investment of computer time.

(2) Rounding errors can cause the corrector to converge to a value different by several units in the last place from the true value (see page 11).

That rounding errors are better controlled by the corrector (advantage (3)) may be seen by writing both predictor and corrector in Lagrangean form as follows, fourth differences being neglected.

Predictor:

$$y_1 - y_0 = \frac{h}{24}(55f_0 - 59f_{-1} + 37f_{-2} - 9f_{-3})$$

Corrector:

$$y_1 - y_0 = \frac{h}{24}(9f_1 + 19f_0 - 5f_{-1} + f_{-2})$$

In the case of the predictor the coefficients are large and alternate in sign, giving a maximum rounding error of 80/24 units. The corresponding figure for the corrector is only 17/24 units.

The method just discussed is not self-starting, since the existence of a series of backward differences is pre-supposed. Also, changing the interval can be non-trivial, especially if interpolation is involved. Most computer subroutines based on this method provide for doubling and halving the interval only; the former is straightforward if sufficient backward values are available, and the latter requires interpolation to halves only.

The methods that follow do not make use of backward values, and are in consequence self-starting. They also permit the interval to be changed by any amount without formality.

Taylor series method

This method is much to be recommended for use in cases in which analytic derivatives of f can be obtained without undue difficulty. y_1 is computed from the Taylor series

$$y_1 = y(x_0+h) = y_0 + hy_0' + \frac{1}{2!}h^2y_0'' + \frac{1}{3!}h^3y_0''' + \dots$$

The method has the very great advantage, compared with finite difference methods, that derivatives are not plagued by rounding errors in the way that differences are. If the terms are all of the same sign, so that rounding errors do not assume large proportions, the interval h is limited only by the radius of convergence of the Taylor series, and can be quite large.

A very useful check may be obtained at small expense in computing time by computing at each step a value of y_{-1} by means of the formula

$$y_{-1} = y(x_0-h) = y_0-hy_0'+\frac{1}{2!}h^2y_0''-\frac{1}{3!}h^3y_0'''+\dots$$

Agreement with the value previously obtained for y_{-1} gives an excellent check both on computational accuracy and on the convergence of the series.

If more than a very few terms are required it is convenient to work in terms of the reduced derivatives

$$Y_p = \frac{h^p}{p!}\frac{d^p y}{dx^p}$$

It is frequently possible to derive by means of Leibnitz theorem a recurrence relation connecting the Y_p; for example, in the case of the equation discussed above, namely $y' = 2xy+1$ it may be shown that

$$(p+1)Y_{p+1} = 2h(xY_p+hY_{p-1})$$

The computer program can be designed so as to compute as many terms as are necessary to make the Taylor series converge; it is not unknown for 20 or 30 terms to be used.

Application of the Taylor series method is at present limited by the necessity for the programmer himself to derive the formulae for the derivatives. Advances in programming languages for symbol manipulation should soon enable this load to be taken from the programmer and put on to the machine.

Runge–Kutta method

Since $$y' = f(x, y)$$

yields $$y'' = \frac{\partial f}{\partial x}+\frac{\partial f}{\partial y}\frac{dy}{dx}$$

the Taylor series may be written

$$y_1 = y_0+hf+\frac{h^2}{2!}(f_x+ff_y)+\dots \tag{1}$$

where f stands for f_0, f_x for $\partial f_0/\partial x$, f_y for $\partial f_0/\partial y$, etc.

The Runge–Kutta method is based on the following equations:

$$\left.\begin{aligned} y_1 &= y_0+k \\ k &= \tfrac{1}{6}(k_0+2k_1+2k_2+k_3) \end{aligned}\right\} \tag{2}$$

where

$$\left.\begin{aligned} k_0 &= hf(x_0, y_0) \\ k_1 &= hf(x_0+\tfrac{1}{2}h, y_0+\tfrac{1}{2}k_0) \\ k_2 &= hf(x_0+\tfrac{1}{2}h, y_0+\tfrac{1}{2}k_1) \\ k_3 &= hf(x_0+h, y_0+k_2) \end{aligned}\right\} \tag{3}$$

If the functions occurring in (3) are expanded by the two-dimensional form of Taylor's series about x_0, y_0—by putting, for example,

$$k_1 = h\Big[f(x_0, y_0)+\left(\frac{h}{2}\frac{\partial}{\partial x}+\frac{k_0}{2}\frac{\partial}{\partial y}\right)f(x_0, y_0) + \frac{1}{2!}\left(\frac{h}{2}\frac{\partial}{\partial x}+\frac{k_0}{2}\frac{\partial}{\partial y}\right)^2 f(x_0, y_0)+\ldots\Big]$$

—and the results substituted in (2), it may be verified after much tedious algebra that (2) agrees with (1) as far as terms in h^4.

Geometrically the Runge–Kutta method may be described in the following terms.

Fig. 1 shows the (x, y) plane. At any point of the plane there is a slope $f(x, y)$.

(1) Draw a straight line OA through (x_0, y_0) with slope $f(x_0, y_0)$. This approximation to the solution of the differential equation gives the increment k_0.

(2) Take the slope corresponding to the mid-point of OA and draw another straight line OB with that slope. This gives the increment k_1.

(3) Draw a third straight line OC with slope corresponding to the mid-point of OB. This gives the increment k_2.

(4) Draw a fourth straight line OD with slope corresponding to the point C. This gives the increment k_3.

The adopted increment k is the weighted mean of k_0, k_1, k_2, k_3 given by (2).

It may be noted that the predictor–corrector and Taylor's series methods give the exact solution (except for rounding errors) when the *solution* is a polynomial. The Runge–Kutta method gives an exact solution (except for rounding errors) when f is a polynomial (of degree not greater than four). Thus, all three methods would give an exact solution (except for rounding errors) for the equation

$$y' = 2(x+1)$$

of which the solution is $\quad y = (x+1)^2$

whereas the Runge–Kutta method would fail to do so for the equation $y' = 2y/(x+1)$ which has the same solution. This observation perhaps accounts for the fact that the Runge–Kutta method enjoyed a reputation for treachery among hand computers who were accustomed to assume that if the solution had smooth differences it was correct.

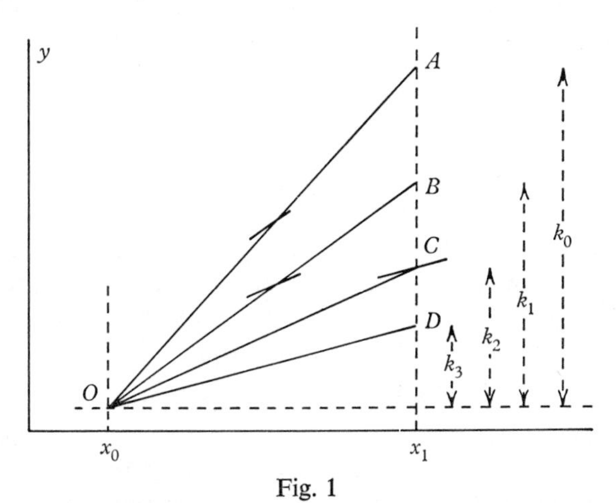

Fig. 1

Although the method just described is the most familiar, there are other methods also going under the name Runge–Kutta. In particular, there is a second-order method, and a variant, due to Gill, of the fourth-order method having certain programming advantages.

Extension to higher-order systems

All the methods discussed can be extended without difficulty to deal with a set of simultaneous differential equations of the form

$$\begin{aligned} y_1' &= f_1(x, y_1, y_2, y_3, \ldots) \\ y_2' &= f_2(x, y_1, y_2, y_3, \ldots) \\ y_3' &= f_3(x, y_1, y_2, y_3, \ldots) \quad \text{etc.} \end{aligned}$$

Moreover, any ordinary differential equations of whatever order can be expressed as a set of first-order equations in many ways. Consider, for example, Bessel's equation of order zero

$$\frac{d^2y}{dx^2}+\frac{1}{x}\frac{dy}{dx}+y = 0$$

This may be written

$$\frac{dy}{dx} = p$$

$$\frac{dp}{dx} = -\frac{p}{x}-y$$

Alternatively, it may be written

$$\frac{dz}{dx} = -xy$$

$$\frac{dy}{dx} = \frac{z}{x}$$

Thus a subroutine for solving a set of simultaneous first-order equations may be used to solve any ordinary differential equation. The exact way in which the equation is expressed as a set of first-order equations is often conditioned by the nature of the starting conditions.

Choice of method

To throw away, as it were, all information about the solution obtained to date, and to start each step of integration *ab initio*, may appear a wasteful proceeding, and the Runge–Kutta method has often been criticized on these grounds. The Taylor series method can be similarly criticized although, where applicable, it has special merits of its own.

The Runge–Kutta method, however, has the virtue that the four values of f used in any one step are all within or on the boundaries of the interval concerned, whereas, in the predictor–corrector method, some of them are well outside and are in consequence stale. In fact, the approximating polynomials used in the latter method stretch over four or five intervals, whereas in the Runge–Kutta method a fourth degree polynomial is fitted to the solution over one interval only. One would thus expect a longer step to be usable. Further, in this method, the ease with which the step can be changed by any amount is a help in keeping the step length near its optimum value over a range of integration.

I consider that for ordinary work the popularity of the Runge–Kutta method is well justified, and that only when computation of derivatives is very time consuming is it worth considering the use of a predictor–

corrector method. Perhaps the main disadvantage of the Runge–Kutta method is that no entirely satisfactory method of obtaining a local indication of when the step length needs changing has so far been devised.

Propagation of errors

The equation $y''-y=0$ has the general solution $Ae^{x}+Be^{-x}$. Suppose initial conditions are chosen with the intention of making $A=0$, and the integration advanced step by step in the direction of x increasing, using, say, the Runge–Kutta method. The integration is started afresh in each interval and new values of A and B are determined for that interval by the new starting conditions. These should be such as to keep $A=0$; inevitably, however, rounding errors will come in, and after a few intervals, A will have a small value different from zero.

The effect of rounding errors is, therefore, to introduce a proportion of the unwanted solution. The effect would, in the above case, eventually be disastrous, since the unwanted solution, which increases with x, will come to swamp the wanted solution which decreases with x. It would be more satisfactory to integrate the equation in the reverse direction (x decreasing) in which case any component of the unwanted solution that were introduced would die away to zero.

The solution e^{-x} is said to be stable when integrating in the direction of x decreasing, but unstable when integrating in the direction of x increasing. This type of instability is a property of the differential equation and its solutions. If possible, one should avoid integrating in an unstable direction. If this cannot be done, one may perhaps be able to work to a degree of accuracy sufficiently high to ensure that by the end of the range of integration the unwanted solution has not grown enough to be troublesome.

Even if the solution itself is stable, it is possible for instability to be introduced by an unfortunate choice of numerical method. This will be investigated using as a model the differential equation

$$y' = Ky$$

where K is constant. In terms of the previous notation we have $f = Ky$ and the true solution is

$$y = y_0 e^{Kx}$$

Thus

$$\begin{aligned} y_1 &= y_0 e^{Kh} \\ &= (1+Kh+\tfrac{1}{2}K^2h^2+\tfrac{1}{6}K^3h^3+\tfrac{1}{24}K^4h^4+\tfrac{1}{120}K^5h^5+\dots)y_0 \end{aligned}$$

Let us solve the equation using the predictor only, i.e. take

$$y_1 = y_0+h(1+\tfrac{1}{2}\nabla+\tfrac{5}{12}\nabla^2+\dots)Ky_0$$

If we neglect the term in ∇ and the following terms, we have

$$y_1 = (1+Kh)y_0$$

This is clearly a genuine approximation to the true solution, albeit a crude one.

Next, let us neglect terms in ∇^2 and higher powers, and express the resulting equation in Lagrangean form. We then have

$$\begin{aligned} y_1 &= y_0 + Kh[y_0 + \tfrac{1}{2}(y_0 - y_{-1})] \\ &= (1+\tfrac{3}{2}Kh)y_0 - \tfrac{1}{2}Khy_{-1} \end{aligned}$$

This is a *difference equation* and may be solved by taking the trial solution $y_{n+1} = \beta y_n$; it is found that β must satisfy

$$\beta^2 - (1+\tfrac{3}{2}Kh)\beta + \tfrac{1}{2}Kh = 0$$

The two solutions of this equation are, when expended in powers of h,

$$\begin{aligned} \beta_0 &= 1+Kh+\tfrac{1}{2}K^2h^2 - \tfrac{1}{4}K^3h^3 + \ldots \\ \beta_1 &= \tfrac{1}{2}Kh(1-Kh+\ldots) \end{aligned}$$

$y_{n+1} = \beta_0 y_n$ is clearly an approximation (as far as h^2) to the true solution. $y_{n+1} = \beta_1 y_n$, however, bears no resemblance to the true solution, and its origin may be questioned. It arises because we have chosen to replace the *first*-order differential equation by a *second*-order difference equation, and it is a solution of the latter but not of the former. Such solutions are sometimes called *parasites.*

In this case, the parasite is harmless, since if h is small, $|\beta_1| \ll |\beta_0|$ and the parasitic solution decays rapidly as the integration is advanced. If further terms of the predictor are taken, additional parasites are introduced, but it may be shown that all decay to zero. Use of the predictor by itself thus always gives stability. When use is also made of a corrector, it is that rather than the predictor that tends to determine whether or not the method is stable. It may be shown that the corrector whose use is illustrated above is always stable.

As an example of an unstable formula we may take a corrector due to Milne; this may be derived as follows:

$$\begin{aligned} y_1 - y_{-1} &= \int_{-1}^{1} f(x, y)\, dx \\ &= \frac{h}{3}(f_{-1} + 4f_0 + f_1) \end{aligned}$$

by application of Simpson's rule. Taking $f = Ky$ and the trial solution

$y_{n+1} = \beta y_n$, we find for the model equation, the following quadratic equation for β

$$\left(1-\frac{Kh}{3}\right)\beta^2-\tfrac{4}{3}Kh\beta-\left(1+\frac{Kh}{3}\right) = 0$$

giving $\quad \beta_0 = 1+Kh+\tfrac{1}{2}K^2h^2+\tfrac{1}{6}K^3h^3+\tfrac{1}{24}K^4h^4+\tfrac{1}{72}K^5h^5+\ldots$

$$\beta_1 = -1+\tfrac{1}{3}Kh+\ldots$$

β_0 gives an excellent approximation to the true solution, and at first sight the method is highly satisfactory. However, the parasite does not in this case decay rapidly, and if $K < 0$ it actually increases whereas the true solution decreases. Reliance on Milne's corrector can therefore lead to difficulties with stability.

The above analysis extends easily to the equation

$$y' = Ky+Lx+M \tag{1}$$

with similar conclusions. Its connection with the general case can be exhibited as follows:

$$\begin{aligned} y' &= f(x, y) \\ &= f_0+(x-x_0)\frac{\partial f_0}{\partial x}+(y-y_0)\frac{\partial f_0}{\partial y} \quad \text{(approx.)} \\ &= \underset{\substack{\uparrow \\ K}}{\frac{\partial f_0}{\partial y}}\, y+\underset{\substack{\uparrow \\ L}}{\frac{\partial f_0}{\partial x}}\, x+\underset{\substack{\uparrow \\ M}}{\left(f_0-x_0\frac{\partial f_0}{\partial x}-y_0\frac{\partial f_0}{\partial y}\right)} \end{aligned}$$

If the coefficients of y, x, and unity can be regarded as approximately constant over a certain range, a stability analysis based on (1) can be applied.

Practical checks on accuracy. The considerations of the last section make it clear that it is useless to rely for an assessment of overall accuracy on an assessment of the error in an individual step, although such an assessment may be needed in a program designed to adjust the step length automatically. A check on overall accuracy is best obtained, as in quadrature, by repeating the entire integration with a smaller interval. If the interval is halved, it is possible to compare the first solution with the second over the whole range of integration, and not just at the end, thus reducing the chance of accidental agreement.

Examples

1. Extend the example on page 53 to $x = \cdot 5$ and $\cdot 6$.

[·5923, ·7671]

2. Show that the following predictor (used without a corrector) is unsatisfactory from the point of view of stability

$$y_1 - y_{-1} = 2hf_0$$

3. Show that the predictor and corrector given in the text are, respectively, in terms of operators

$$y_1 - y_0 = -h\{\nabla/[(1-\nabla)\log(1-\nabla)]\}f_0$$
$$y_1 - y_0 = -h[\nabla/\log(1-\nabla)]f_1$$

and hence obtain the expansions given.

4. Solve the equation $y' = 2xy+1$, given that $y = 0$ at $x = 0$, by the use of the Taylor series, taking $h = \cdot 2$ and going as far as $x = 1 \cdot 0$.

5. A solution is required to the differential equation

$$y' = x^2 - y^2$$

for which $y = 1$ when $x = 0$.

Obtain the following recurrence relations for the reduced derivatives

$$Y_p = \frac{h^p}{p!}\frac{d^p y}{dx^p}$$

$$(p+1)\,Y_{p+1} = -h\sum_{r=0}^{p} Y_r Y_{p-r} \quad (p > 2)$$

Use this to obtain the solution required taking $h = \cdot 1$, and going as far as $x = \cdot 5$. (This example can be worked on a desk machine, keeping 5 decimals, or it can be programmed for a digital computer.)

6

PROBLEMS REDUCIBLE TO SIMULTANEOUS EQUATIONS

Two-point boundary problems

The methods of the last chapter are applicable to sets of ordinary differential equations in which all the boundary conditions apply at one point, which can be taken as the starting point for the integration. We have what is known as a marching problem. Such methods would not be applicable directly to the solution of the differential equation

$$y'' = x^2y-1 \tag{1}$$

given that $y = 1$ when $x = 0$

and $y = 2$ when $x = 5$

They can be applied indirectly, using a trial and error method in which the integration is started from one end with a trial value for y' which is adjusted systematically until a solution is obtained which satisfies the condition at the second point. This method is perfectly workable, and can be performed automatically in a digital computer; it is often useful for non-linear differential equations, but in the linear case it is usually better to proceed otherwise.

Let us insert in (1) the crudest possible approximation for y'', namely

$$y''_r = \frac{\delta^2 y_r}{h^2} = \frac{y_{r-1}-2y_r+y_{r+1}}{h^2}$$

This gives
$$y_{r-1}-(2+h^2x_r^2)y_r+y_{r+1}+h^2 = 0 \tag{2}$$

If we take $h = 1$ there are four unknown values of y in all, namely y_1, y_2, y_3, y_4. In addition, the boundary conditions give $y_0 = 1$, $y_5 = 2$. Equation (2) holds at each internal point, and by taking $r = 1, 2, 3, 4$ in sequence we obtain the following equations:

$$\begin{aligned}
\underline{1}-3y_1+y_2 \qquad\qquad\qquad &+1 = 0\\
y_1-6y_2+y_3 \qquad\quad &+1 = 0\\
y_2-11y_3+y_4 \quad &+1 = 0\\
y_3-18y_4+\underline{2} \; &+1 = 0
\end{aligned}$$

The differential equation is thus represented as a set of simultaneous algebraic equations. It is to be noted that the boundary conditions are included in this representation (the underlined figures come from the boundary conditions).

The above equations may conveniently be expressed in matrix form as follows:

$$\begin{bmatrix} -3 & 1 & & \\ 1 & -6 & 1 & \\ & 1 & -11 & 1 \\ & & 1 & -18 \end{bmatrix} \begin{bmatrix} y_1 \\ y_2 \\ y_3 \\ y_4 \end{bmatrix} = -\begin{bmatrix} 2 \\ 1 \\ 1 \\ 3 \end{bmatrix}$$

In the matrix on the left, all terms distant more than one place from the diagonal are zero, and the zeros have been omitted. It is an example of a *banded* matrix.

A slightly more complex situation arises if, instead of the function, the derivative is given at one boundary point; for example, if we have

$$y' = 0 \quad \text{when} \quad x = 0$$

There is now an extra unknown y_0, and an extra equation which expresses the known fact about the derivative. By differentiating the Newton–Gregory formula with respect to θ and putting $\theta = 0$, it is easily shown that

$$hy'_0 = (\Delta - \tfrac{1}{2}\Delta^2 + \tfrac{1}{3}\Delta^3 - \dots)y_0$$

The crudest approximation to $y'_0 = 0$, obtained by neglecting Δ^2 and higher powers is

$$y_0 - y_1 = 0$$

A better approximation, neglecting Δ^3, is

$$y_1 - y_0 - \tfrac{1}{2}(y_0 - 2y_1 + y_2) = 0$$

or

$$\tfrac{3}{2}y_0 - 2y_1 + \tfrac{1}{2}y_2 = 0$$

Using the latter approximation, we have for the matrix equation

$$\begin{bmatrix} \frac{3}{2} & -2 & \frac{1}{2} & & \\ 1 & -3 & 1 & & \\ & 1 & -6 & 1 & \\ & & 1 & -11 & 1 \\ & & & 1 & -18 \end{bmatrix} \begin{bmatrix} y_0 \\ y_1 \\ y_2 \\ y_3 \\ y_4 \end{bmatrix} = -\begin{bmatrix} 0 \\ 1 \\ 1 \\ 1 \\ 3 \end{bmatrix}$$

The convergence of the forward difference formula used for the derivative is not good, and a better approach in situations of this type is to make use of points outside the range within which the solutions are required. Such points are known as *fictitious* points, although they are in no way fictitious as regards the equation, only as regards the given problem.

If we take a fictitious point y_{-1}, we have, as an approximation,

$$2hy_0' = y_1 - y_{-1}$$

giving for $y_0' = 0$ $$y_{-1} - y_1 = 0$$

We must also satisfy the differential equation at $x = 0$; by putting $r = 0$ in (2) we obtain

$$y_{-1} - 2y_0 + y_1 + 1 = 0$$

The matrix equation is now

$$\begin{bmatrix} 1 & 0 & -1 & & & \\ 1 & -2 & 1 & & & \\ & 1 & -3 & 1 & & \\ & & 1 & -6 & 1 & \\ & & & 1 & -11 & 1 \\ & & & & 1 & -18 \end{bmatrix} \begin{bmatrix} y_{-1} \\ y_0 \\ y_1 \\ y_2 \\ y_3 \\ y_4 \end{bmatrix} = - \begin{bmatrix} 0 \\ 1 \\ 1 \\ 1 \\ 1 \\ 3 \end{bmatrix}$$

Similar methods may be used to deal with the more complicated case in which the boundary condition is of the form $ay' + by = c$. When the derivative alone occurs in the boundary condition, and there is no particular reason why the boundary point should be one of the points at which the function is computed, it may be better to choose the points x_{-1}, x_0, x_1, etc. so that the boundary point is mid-way between x_{-1} and x_0. Use can then be made of the following relatively accurate formula for the derivative:

$$hy_{\frac{1}{2}}' = y_0 - y_{-1}$$

Higher-order operators. It is possible to make use of more accurate finite difference representations of the differential operator than that used above. For simplicity, the discussion will be confined to the case in which function values (rather than derivatives) are given at the two boundary points.

At first sight there appears to be no difficulty in writing

$$\begin{aligned} h^2 y_r'' &= \delta^2 y_r - \tfrac{1}{12}\delta^4 y_r \\ &= \tfrac{1}{12}(-y_{r-2} + 16y_{r-1} - 30y_r + 16y_{r+1} - y_{r+2}) \qquad (3) \end{aligned}$$

instead of the expression used earlier in which $\delta^4 y_r$ was omitted. A difficulty arises, however, at points adjacent to the boundary, since either y_{r-2} or y_{r+2} then lies outside the range of integration. Once again use may be made of fictitious points. For example, if in (3) we take $r = 1$, we have

$$h^2y_1'' = \tfrac{1}{12}(-y_{-1}+16y_0-30y_1+16y_2-y_3)$$

where y_{-1} corresponds to a fictitious point. In this case, the coefficient associated with the fictitious point is small, and a crude approximation to y_{-1} suffices. This may conveniently be obtained from the formula

$$h^2y_r'' \doteq \delta^2 y_r = y_{r-1}-2y_r+y_{r+1}$$

applied at the point $r = 0$, that is to say

$$h^2y_0'' = y_{-1}-2y_0+y_1$$

where y_{-1} is regarded as the unknown.

Instead of using fictitious points, there is the option of using, at points adjacent to the boundary, a finite difference formula of the correct order based on forward or backward differences. This may have to be specially derived. In the case just considered the following formula, obtained by differentiating the Newton–Gregory formula twice with respect to θ, and putting $\theta = 1$, is available

$$h^2y_1'' = \Delta^2y_0-\tfrac{1}{12}\Delta^4y_0+\tfrac{1}{12}\Delta^5y_0-\tfrac{13}{180}\Delta^6y_0+\ldots$$

Provided that the solution of the differential equation is well-behaved beyond the boundary, the use of fictitious points is generally to be preferred on account of the better convergence of the finite difference series. This is particularly true in the more difficult case in which derivatives occur in the boundary conditions.

Note that, when the more accurate finite difference form of the differential operator is used, the matrix still consists of a band of non-zero elements bordering the principle diagonal. The band is, however, broader than it was when the simpler approximation was used.

Two point boundary problems in ordinary differential equations thus lead to the problem of solving sets of simultaneous algebraic equations. In the linear case these may be solved by the methods outlined in the next chapter. Non-linear equations must in general be solved iteratively, each case being treated on its merits.

Note that there are two distinct sources of error: (1) truncation error in the finite-difference approximations, and (2) error in the solution of the algebraic equations. There is no point in solving the equations to 10 decimal places if the finite-difference approximations are only good to 5 decimals.

A check of overall error is best obtained by repeating the whole solution with a smaller interval; this covers truncation error, and also errors in the solution of the simultaneous equations. Attempts to estimate truncation error by examining neglected terms in the finite difference approximations to the differential operators encounter the difficulty that there is no simple way of relating residuals in equations to errors in the solution (see p. 75).

Poisson's equation

An example of a partial differential equation that gives rise to a set of linear equations is Poisson's equation

$$\frac{\partial^2 f(x, y)}{\partial x^2} + \frac{\partial^2 f(x, y)}{\partial y^2} = g(x, y) \qquad (1)$$

where f is given on a closed boundary, which we will take, for simplicity, to be rectangular. We replace the continuous functions $f(x, y)$ and $g(x, y)$ by functions known only at the intersections of a rectangular grid. Let

$$f_{r,s} = f(x_r, y_s)$$

where (x_r, y_s) is one of the intersections. Similarly,

$$g_{r,s} = g(x_r, y_s)$$

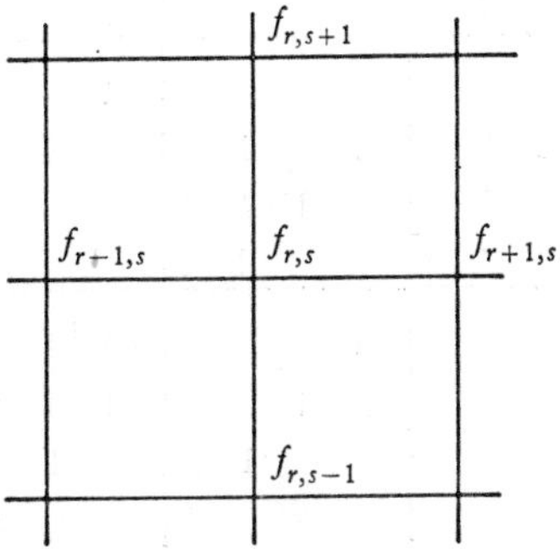

Fig. 1

In order to express the partial derivatives, differencing in two directions is necessary, and we may write as a finite difference version of (1)

$$\frac{\delta_x^2 f_{r,s}}{h^2} + \frac{\delta_y^2 f_{r,s}}{h^2} = g_{r,s}$$

or, in Lagrangean form,

$$f_{r,s+1} + f_{r,s-1} + f_{r+1,s} + f_{r-1,s} - 4f_{r,s} = h^2 g_{r,s}$$

This may be re-written as

$$\begin{bmatrix} & f_{r,s+1} & \\ +f_{r-1,s} & -4f_{r,s} & +f_{r+1,s} \\ & +f_{r,s-1} & \end{bmatrix} = h^2 g_{r,s}$$

where the terms have been displayed in such a way that their positions correspond to the positions of the corresponding intersections of the grid. This idea may be carried further, and the coefficients only of the f's displayed as follows:

$$\begin{bmatrix} & 1 & \\ 1 & -4 & 1 \\ & 1 & \end{bmatrix} f_{r,s} = h^2 g_{r,s}$$

The contents of the square brackets form an operator or template that can be placed anywhere on the grid to generate one of the set of difference equations representing the differential equation. For many purposes this is the best form in which to handle the equations. It is

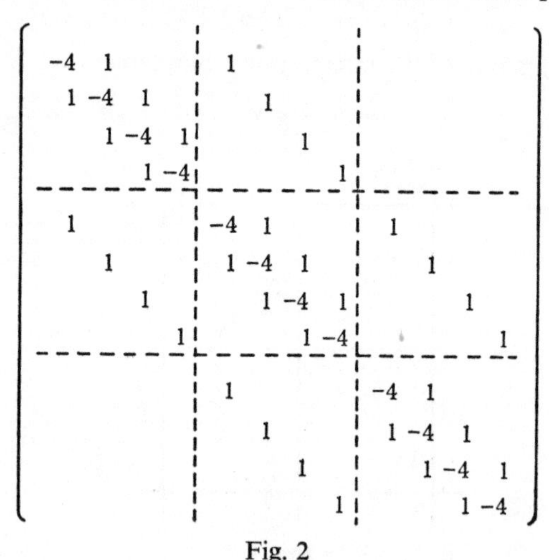

Fig. 2

not necessary to write the equations out explicitly since they all have the same form. The above operator can be applied at points adjacent to the boundary, but values of f corresponding to points actually on the boundary will, of course, appear in the equations as constants, not as unknowns.

The equations can also be expressed in matrix form. If the grid has n by m points, the matrix of the coefficients will have nm rows and columns. An example with $n = 3$ and $m = 4$ is shown in Fig. 2. It has natural partitioning into submatrices as indicated by the dotted lines. Note that the matrix has a banded form.

A disadvantage of the matrix representation is that the symmetry between x and y is lost. Points that are adjacent on the grid are close together in the matrix only if they are on the same row; if they are on different rows they are wide apart.

Fictitious points outside the region of integration may be used for dealing with boundary conditions involving derivatives. The formulation follows closely that already given for ordinary differential equations.

Higher-order operators. If the term in δ^4 is retained in the representation of f'', instead of the operator

$$\begin{bmatrix} & 1 & \\ 1 & -4 & 1 \\ & 1 & \end{bmatrix} \quad \text{we obtain} \quad \begin{bmatrix} & & -1 & & \\ & & 16 & & \\ -1 & 16 & -60 & 16 & -1 \\ & & 16 & & \\ & & -1 & & \end{bmatrix}$$

In order that this operator may be used near the boundary it is again convenient to make use of fictitious points outside the area over which the solution is required. Approximate values only are required at the fictitious points, and these may conveniently be obtained by the application of the simpler operator to a point on the boundary, the value of f at the fictitious point being regarded as an unknown. Again the discussion given for ordinary differential equations may be extended to this case.

Examples

1. Express in matrix form a discrete approximation to the differential equation $y'' = xy$ given that

$$y'+y = 1 \quad \text{at} \quad x = 0$$

$$y = 1 \quad \text{at} \quad x = 1$$

and taking $h = \cdot 2$.

2. Derive a suitable operator for solving the following differential equation on a rectangular grid

$$\frac{\partial^2 f}{\partial x^2}+\frac{\partial^2 f}{\partial y^2}+\frac{1}{x}\frac{\partial f}{\partial x} = 0$$

7

SOLUTION OF LINEAR ALGEBRAIC EQUATIONS

Methods for solving linear equations can be divided into *direct* methods, which are equivalent to elimination, and *indirect* or iterative methods. Direct methods are generally to be preferred, except in the following special circumstances when indirect methods are indicated:

(1) the number of equations is large in relation to the digital computer available,
(2) the equations are such that the convergence of a suitably chosen iterative method is specially rapid,
(3) a specially good starting approximation is available.

The number of equations that can be handled by direct methods has increased steadily with the increasing power of digital computers. With a modern computer of reasonable power (say a multiplication time of 250 μs, and at least 16,000 words of core storage) it takes between one and 2 min to solve a set of 100 equations in 100 unknowns. The time increases with the number of equations n by a factor between n^3 and n^4. For a given computer this rule breaks down when n becomes so large that all the coefficients cannot be accommodated at the same time in the high speed store, so that an auxiliary store with longer access time has to be used.

The above remarks refer to equations of general form. Banded equations such as arise from differential equations can be handled in much larger sets—up to several thousand in the computer mentioned above—and the time for solution increases more or less linearly with n.

Triangular form. There is one case in which solution of the equation

$$Ax = b$$

where A is a square matrix, and x and b are column vectors, becomes trivial, namely when all the elements of A above the diagonal are zero; for example, in the case

$$\begin{aligned} x_1 \qquad\qquad &= 3 \\ 2x_1+4x_2 \qquad &= 2 \\ 3x_1+5x_2+6x_3 &= 1 \end{aligned}$$

The first equation gives x_1 explicitly, and the other variables are then obtainable in order from the remaining equations. A similar situation would exist if all the elements of A *below* the diagonal were zero.

Matrices with zeros above or below the diagonal are said to be *lower triangular* or *upper triangular* respectively, and the process of solving a set of equations whose matrix is of one of these forms is sometimes called *back substitution.*

Pivotal condensation. This is a name given to the solution of linear equations by elimination systematically performed. Consider the equations

$$3x_1+3x_2+2x_3 = 7 \tag{1}$$

$$18x_1+12x_2+6x_3 = 1 \tag{2}$$

$$27x_1+6x_2+18x_3 = 2 \tag{3}$$

x_1 may be eliminated from all but (1) as follows:

Multiply (1) by 18/3 and subtract from (2)

Multiply (1) by 27/3 and subtract from (3)

x_2 is similarly eliminated from the new third equation by subtracting a multiple of the new second equation. These operations have the effect of converting the equations to the following triangular form:

$$\begin{aligned} 3x_1+3x_2+2x_3 &= 7 \\ -6x_2-6x_3 &= -41 \\ 21x_3 &= 165/2 \end{aligned}$$

The coefficient by which the other coefficients are divided in performing the elimination is known as the pivot. In the above example, the pivot is the coefficient—3—of the first term of the first equation.

The process as described is quite satisfactory if the arithmetic can be performed exactly, but in general the effect of multiplying by factors such as 18/3 and 27/3, which are greater than unity, is to increase the effect of rounding errors. This trouble arises from an unfortunate choice of pivot, and can be avoided by choosing as pivot the largest coefficient in the column concerned. If this is done it is never necessary to multiply by a number greater than unity. It is convenient to effect, if necessary, an interchange of rows so as to leave the final matrix in triangular form.

An alternative way of describing the elimination procedure described above is to say that both sides of the original equation, regarded as a vector equation, have been multiplied by a certain (triangular) matrix.

Viewed from this point of view, pivotal condensation is a means of finding a lower triangular matrix L such that LA is upper triangular; solution of the matrix equation

$$LAx = Lb$$

is then effected by back substitution.

It may easily be verified that in the above example we have

$$L = \begin{bmatrix} 1 & 0 & 0 \\ -6 & 1 & 0 \\ 12 & -\frac{7}{2} & 1 \end{bmatrix}$$

Triangular decomposition. In this method the matrix A is expressed as the product of a lower triangular matrix L and an upper triangular matrix U, that is

$$LU = A$$

The diagonal terms of L or U may be specified arbitrarily, and it is convenient to make those of L equal to one.

The manner in which the decomposition is made will be clear from the following example in which A has three rows and columns.

$$\begin{bmatrix} 1 & 0 & 0 \\ l_{21} & 1 & 0 \\ l_{31} & l_{32} & 1 \end{bmatrix} \begin{bmatrix} u_{11} & u_{12} & u_{13} \\ 0 & u_{22} & u_{23} \\ 0 & 0 & u_{33} \end{bmatrix} = \begin{bmatrix} a_{11} & a_{12} & a_{13} \\ a_{21} & a_{22} & a_{23} \\ a_{31} & a_{32} & a_{33} \end{bmatrix}$$

Multiplied out these become

$$\begin{aligned}
u_{11} &= a_{11} \\
u_{12} &= a_{12} \\
u_{13} &= a_{13} \\
l_{21}u_{11} &= a_{21} \\
l_{21}u_{12}+u_{22} &= a_{22} \\
l_{21}u_{13}+u_{23} &= a_{23} \\
l_{31}u_{11} &= a_{31} \\
l_{31}u_{12}+l_{32}u_{22} &= a_{32} \\
l_{31}u_{13}+l_{32}u_{23}+u_{33} &= a_{33}
\end{aligned}$$

It may be verified that the unknown components of L and U can be evaluated in a straightforward manner from these equations used in the order in which they are given.

Once A has been decomposed into LU, the solution may be effected by noting that

$$LUx = b$$

is equivalent to

$$Ux = y$$
$$Ly = b$$

The second equation is solved by back substitution, giving y; the first is then solved by another back substitution, giving x.

Gauss–Seidel method. This is an indirect method. Suppose x_1, x_2, x_3 is an approximation to the solution of a set of equations, and that we have

$$a_{11}x_1+a_{12}x_2+a_{13}x_3+\ldots-b_1 = R_1 \quad (1)$$
$$a_{21}x_1+a_{22}x_2+a_{23}x_3+\ldots-b_2 = R_2 \quad (2)$$
$$a_{31}x_1+a_{32}x_2+a_{33}x_3+\ldots-b_3 = R_3 \quad (3)$$
$$\ldots$$

R_1, R_2, R_3 are called residuals, and the object is to derive improved values of x_1, x_2, x_3 for which the residuals are as small as desired. We proceed as follows:

With the aid of (1) choose a new x_1 so as to make $R_1 = 0$.
With the aid of (2) choose a new x_2 so as to make $R_2 = 0$ upsetting R_1.
With the aid of (3) choose a new x_3 so as to make $R_3 = 0$ upsetting R_2.
Continue until all the residuals in turn have been (temporarily) reduced to zero, and then repeat.

It depends on the matrix A whether or not the process converges in the sense that the residuals can be made as small as desired. An important case in which one would expect convergence to occur—and the expectation is borne out in practice—is when A has a strong leading diagonal, i.e. when the terms on the diagonal are numerically large compared with the other terms. Matrices arising from differential equations are normally of this type. Convergence always occurs when A is positive definite, and this may be proved.

The method is subject to many variants. For example, the equations may be taken in some different order, one rule being always to take the equation with the largest residual. With some equations, it may be a good strategy not to reduce the residuals to zero, but to cause them to be multiplied by some quantity α such that $|\alpha| < 1$.

Relaxation. Southwell developed a method of the type described above to a state in which it formed a powerful tool in the hands of a skilled worker. He approached the subject by way of mechanical frameworks, in which he imagined applied constraints being systematically relaxed. This gave the name *relaxation* by which the method became known. It proved particularly useful for solving the large sets of equations arising from the discretization of partial differential equations of the elliptic type (e.g. Poisson's equation). Here Southwell provided a convenient way of arranging the work on a large sheet of paper ruled into a grid.

Relaxation is a method in which the skill of the operator in choosing where and by what amount to make changes in the trial solution is at a premium. Generally he concentrates attention on points where the residuals are greatest, and in the case of Poisson-like equations it always pays to 'over-relax', i.e. multiply the residuals by some factor, α, lying between 0 and -1.

Methods based on an examination of the size of the residuals are unattractive in a digital computer, since it takes as long to examine a residual to see how large it is as it does to reduce the residual to zero or to multiply it by a factor α. Consequently, methods are used in which the various grid points are operated on in some fixed order. A very common procedure is to take them row by row or column by column, using always the most recently obtained values of the function. This method has a certain lopsidedness since, at a typical point, two of the surrounding points will have been recently visited on the same sweep, whereas two will not have been visited since the last sweep. This can be avoided if the grid is divided into two subgrids, as shown in Fig. 1, where the crosses indicate points belonging to one subgrid and the circles indicate points belonging to the other subgrid. The sweep is then done in two parts, one subgrid being first covered, and then the other.

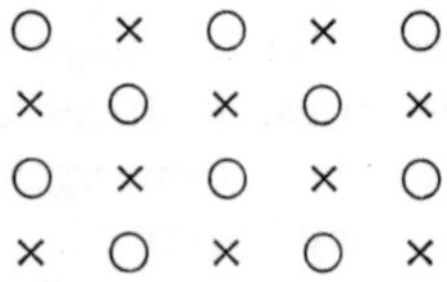

```
O × O × O
× O × O ×
O × O × O
× O × O ×
```

Fig. 1

Methods in which the points are swept systematically lose a factor of perhaps 10 as compared with relaxation performed by a human operator, but, nevertheless, the greater speed of the machine gives a greatly increased overall speed. For Poisson-type equations using a large number of points in the grid it is found that a value of α quite near to -1 is usually required to give optimum convergence.

Errors and residuals. It may be asked how far an iteration to reduce residuals should be carried in order to achieve specified accuracy in the unknowns. One answer is that the process should be continued until the approximate solution, to the accuracy considered, shows no further change. This perhaps settles the matter if the coefficients of the equations are given exactly (e.g. are integers), but if they are rounded numbers, there is clearly no point in reducing the residuals beyond the point at which they become dominated by the rounding errors of the coefficients. In such a case there is a limit to the accuracy with which the unknowns are determined by the equations in their rounded form. Unfortunately, there is no simple way of determining what this limit is.

As an extreme example, consider the equations:

$$x+2y \quad = 4$$
$$x+2{\cdot}002y = 4{\cdot}003$$

The trial solution $x = 0$, $y = 2$ gives residuals 0 and ·001 and might be deemed a good approximation; however, the very different trial solution $x = 2$, $y = 1$ which gives residuals 0 and $-{\cdot}001$ is just as good judged by the smallness of the residuals. $x = 1$, $y = \frac{3}{2}$ gives zero residuals, and may be said to be the true solution. This last statement is only meaningful, however, if the coefficients are given exactly; if the coefficients 2·002 and 4·003 are rounded-off approximations to numbers which cannot be accurately expressed to three decimal places, then the truth is that the equations define the unknowns with hardly any accuracy at all. Such equations are said to be *ill-conditioned*, and are a plague to practitioners in the art of computing. One of the signs of ill-condition is that the determinant is small.

Example

Solve the set of four equations given on page 63 by a Gauss–Seidel iterative process:

(1) taking the equations in order and reducing residuals successively to zero ($\alpha = 0$);

(2) as (1) but selecting at each stage the equation with the largest residual;

(3) as (2) but with $\alpha = -{\cdot}25$.

(This example can be worked by hand, but students who have access to a digital computer will no doubt prefer to write a program. Given a digital computer it would be more instructive to take the larger set of equations obtained by putting $h = {\cdot}2$ in equation (2) on page 63.)

FURTHER READING

For further study *Modern Computing Methods* written by the staff of the Mathematics Division of the National Physical Laboratory and published by H.M.S.O. is much to be recommended. For reference purposes, all students of numerical analysis should have access to *Introduction to Numerical Analysis* by F. B. Hildebrand (McGraw-Hill).

INDEX